Beyond Solitude

The Science and Mystery of Cosmic Civilisations

Clara Bennett

Global East-West (GEW) London

Collection: Exploring the Unseen.

Global East-West (GEW) London.

Contents

Also by Clara Bennett

In the collection: Exploring the Unseen

- **Parapsychology:** The Science Of The Paranormal. GEW. January 2025.
- **Alien Chronicles**: Unraveling the Mysteries of Extraterrestrial Visitations. GEW. October 2024.

Part One

Fermi's Children: Searching for Intelligence in a Silent Galaxy

Introduction

The Question That Will Not Die

Since the time we started looking up at the night sky, we have been wondering whether we are alone. This is the question, perhaps the greatest question we can put, which resonates across millennia of human thought. It is carved on the walls of ancient caves of deities, it is woven in the mythology of the creation of Sumer and Egypt, it is discussed in the philosophical schools of Greece.

Anaxagoras, in the 5th century BCE, speculated that the Moon and stars were other worlds with inhabitants. Centuries later, the Roman poet Lucretius wrote in *De Rerum Natura*, "Nothing in the universe is unique and alone... therefore we must realise that there are other worlds in other regions of the sky, and different races of men and different species of beasts." This was not science as we know it, but it was a powerful intuition: that the cosmos, in its vast regularity, would not craft a solitary masterpiece.

For much of human history, this intuition existed in a realm of pure speculation, forever beyond the reach of verification. The turning point, the moment the question began to shed its mythological skin for a scientific one, arrived with Nicolaus Copernicus. By displacing Earth from the centre of the universe, he initiated a revolution of perspective. We were not the focal point of creation. If our world was but one planet orbiting an ordinary star, then surely those myriad other points of light could host their own worlds, their own stories? Giordano Bruno, burnt at the stake in 1600, championed this plurality of worlds with fervent, if unsupported, conviction. With the invention of the telescope, Galileo Galilei brought Jupiter's moons into focus—other worlds, indeed, orbiting another centre. The cosmic stage was set; humanity just lacked the script.

The 20th century transformed the search from a philosophical exercise into a quantifiable scientific endeavour. In 1950, over a casual lunch at Los Alamos, physicist Enrico Fermi posed his famous, devastatingly simple query: "But where is everybody?" If the universe is so old and so vast, he reasoned, then surely intelligent life should have arisen elsewhere long ago, and its presence should by now be obvious. The silence we perceived was a paradox. A decade later, astronomer Frank Drake attempted to formalise the problem. His eponymous equation, while not providing an answer, provided a framework: a chain of probabilities concerning star formation, habitable planets, the origin and evolution of life, and the lifespan of civilisations. It gave scientists a checklist for investigation.

Since then, the floodgates have opened. Missions like NASA's Kepler Space Telescope, which stared unblinkingly at a patch of sky, and its successor, TESS (the Transiting Exoplanet Survey Satellite), have turned a trickle of discoveries into a torrent. As of the mid-2020s, we have confirmed over 5,500 exoplanets. Statistically, extrapolating from these missions, astronomers estimate our Milky Way alone contains *billions* of planets, with a significant fraction residing in the 'Goldilocks Zone'—the orbital band around a star where temperatures could allow liquid water to exist on a rocky surface. Water, carbon chemistry, and a stable energy source: the cosmic recipe for life, as we understand it, appears to be written in surprisingly common ingredients.

This is why the year 2026, and the decade it inaugurates, represents a pivotal moment in a story thousands of years in the making. We are no longer simply wondering. We are *looking* with tools of unprecedented power and precision. A new generation of observatories is coming online, fundamentally shifting our strategy from simply finding planets to intimately studying them. The James Webb Space Telescope (JWST), already operational, possesses the extraordinary capability to analyse the atmospheric composition of distant worlds. It can search for 'biosignatures'—chemical imbalances, like the co-presence of oxygen and methane, that on Earth are sustained by life.

On the ground, the Vera C. Rubin Observatory in Chile will soon begin the Legacy Survey of Space and Time (LSST), a ten-year cinematic panorama of the sky that will detect millions of transient events and subtle changes. It could, by sheer statistical brute force, spot the tell-tale dimming of a star caused by a massive artificial structure or catch other anomalies that defy natural explanation. Furthermore, next-generation radio telescope arrays, like the Square Kilometre Array (SKA), will scan the cosmic radio whisper with a sensitivity that dwarfs all previous efforts, capable of detecting the equivalent of a terrestrial airport radar from a star dozens of light-years away.

We are, for the first time in history, building the tools that could plausibly answer the ancient question. We are moving from the astronomy of *where* planets are to the astrobiology of *what* they are like. Within the next decade, we may have our first credible, peer-reviewed evidence of a potential biosignature in the atmosphere of an exoplanet. We may identify a candidate technosignature—a sign of technology—that withstands initial scrutiny. Or, we may witness the 'Great Silence' grow deeper and more perplexing as our searches, now capable of detecting the signals we hypothesise, continue to find nothing but natural static.

What is no longer a question that cannot die is now a question that can be actively followed by science. We are no longer mere star peepers; we are galactic detectives, filtering photons and data to give us leads to the biggest mystery of the universe. It is the experience of this process of the development of ancient myth to the contemporary mission that has led us to the periphery of the possible revelation. It is time to start by asking ourselves how we got where we are, and what the next phase in our great inquiry of the universe may turn out to be.

1

The Cosmic Numbers Game

The night sky has been the canvas of human imagination since millennia. We have filled it with gods and monsters, heroes and beasts, and we have seen our own histories in the constellations. But within the final slice of cosmic time--a half-century, sixty years, at most--we have experienced a revolution of this viewpoint as radical as anything we have ever known. We no longer fantasise about other worlds, but have started discovering them. It is no longer a question of whether there are other planets, but how many there could be, and how many of them could harbour life, and perhaps, minds similar to our own. To wrestle with this transformation of myth to mathematics, we should start with a simple, beautiful, and suggestively powerful formula: the Drake Equation.

The equation was originally developed in 1961 by the forward-looking astronomer Frank Drake as part of the preparations of the first SETI meeting, and was never meant to give a single, definitive answer. Its genius is in its form; it dissects the great, threatening question, (Are we alone?) to a sequence of smaller, more manageable probabilistic steps. Written as $N = R^* \cdot f_p \cdot n_e \cdot f_l \cdot f_i \cdot f_c \cdot L$ serves as a conceptual framework for our ignorance, a map of what we need to know.

Let's unpack its terms. N represents the number of communicative civilisations in our galaxy with whom we might potentially connect. To arrive at N, we start with R^*, the average rate of star formation in the Milky Way. This is a number we can measure with some confidence: roughly 1.5 to 3 new stars per year. We then multiply by f_p, the fraction of those stars that have planetary systems. Just a few decades ago, this was pure speculation. Today, thanks to missions like NASA's Kepler and TESS, we know planets are not the exception but the rule. It appears f_p is very close to 1; almost every star is likely to host at least one world.

The next factor, n_e, asks a more subtle question: of the planets orbiting a star, how many are in the "habitable zone"? This is the region where conditions might allow for liquid water to exist on a planet's surface—not too hot, not too cold. It's a useful, if simplistic, starting point, a proxy for Earth-like conditions. Kepler's data suggests that roughly one in five Sun-like stars has an Earth-sized planet in its habitable zone. When we apply this to the hundreds of billions of stars in our galaxy, the numbers become staggering. There could be *tens of billions* of potentially habitable, rocky worlds in the Milky Way alone.

Assuming life does arise, f_i represents the fraction of those life-bearing planets where intelligence evolves. Intelligence, as we define it, is not an inevitable endpoint of evolution. For over three billion years, life on Earth was content with microbial existence. Complex, multicellular life is a relatively recent innovation, and human-like intelligence has only flashed into existence once. Is intelligence a rare fluke, a quirky adaptation like the elephant's trunk, or is it a convergent feature that evolution stumbles upon repeatedly given enough time? We simply do not know.

Then comes f_c: the fraction of intelligent species that develop technology that releases detectable signs of their existence into space—technosignatures. This could be radio transmissions, laser pulses, massive engineering projects, or atmospheric pollution. A civilisation might be contemplative and inward-focused, or it might transcend biological form before developing radio telescopes. Our own technological phase is, so far, barely a century old.

Finally, and perhaps most critically, is L: the average lifetime of such a communicative, technological civilisation. This is the great unknown upon which the entire equation pivots. Does L last for a few turbulent centuries before a species succumbs to nuclear war, ecological collapse, or a pandemic? Or can a civilisation achieve stability, perhaps merging with its technology, and persist for millions or even billions of years? If L is large, then even if the other probabilities are small, the galaxy should be teeming with ancient, enduring intelligence. If L is tragically short, the cosmos may be littered with the silent tombs of species that briefly flickered and died.

The modern power of the Drake Equation lies in our newfound ability to constrain its early terms. We now know that the galaxy is abundantly supplied with planets, and habitable real estate is common. This has shifted the weight of the puzzle onto the latter terms, particularly f_l, f_i, and L. The "cosmic numbers game" is thus increasingly a game of biological and sociological probability.

Central to this game is the concept of the habitable zone, or the "Goldilocks zone". Early calculations were simple, based on a star's luminosity and a planet's distance. But we now understand that planetary habitability is a far more nuanced tapestry. A planet's atmosphere is its blanket and shield; a thick greenhouse atmosphere like Venus's can render a world in the nominal habitable zone a hellscape, while a sufficiently insulating atmosphere could warm a distant moon like Titan. Geological activity, which drives the carbon-silicate cycle and helps regulate climate over aeons, is another critical factor. So too is the presence of a large moon to stabilise axial tilt or a magnetic field to deflect stellar radiation. The "Rare Earth" hypothesis argues that the confluence of all these factors necessary for complex animal life may be exceedingly unusual. Others counter that life is ingenious and tenacious, and we should expect to find it in the subsurface oceans of icy moons, in the clouds of gas giants, or in forms we have not yet dreamed of.

When we plug our modern knowledge into the Drake Equation, even with conservative estimates, the result is provocative. With hundreds of billions of stars, most with planets, and billions of those in habitable zones, it seems statistically almost absurd if life emerged only once. The numbers whisper that the galaxy should be alive. And if even a tiny fraction of life-bearing worlds give rise to intelligence and technology, then the number N should be significant.

Yet, when we listen, we hear silence. When we look, we see no unambiguous signs of cosmic engineering. This is the tension at the heart of our search—the chasm between the encouraging probabilities suggested by the cosmic numbers and the austere reality of the "Great Silence". The Drake Equation does not solve the puzzle of our solitude; it precisely defines its parameters. It tells us that the answer lies not in a scarcity of worlds, but in the probabilities of life's emergence, intelligence's flowering, and, most hauntingly, a civilisation's longevity. As we move from counting planets to probing their atmospheres for biosignatures like oxygen and methane, we are beginning to test the term f_l not with philosophy, but with data. The cosmic numbers game is now being played on a board we are only just beginning to see, and the next move promises to be the most revealing in history.

References For Further Reading

The following references span the key topics covered in this chapter, including the Drake Equation, habitable zones, biosignatures, technosignatures, and the Fermi Paradox.

1. Frank, A., & Sullivan, W. T. (2015). A new empirical constraint on the prevalence of technological species in the universe. Astrobiology, 16(5), 359–362. https://doi.org/10.1089/ast.2015.1418

This paper provides updated empirical constraints on the Drake Equation using modern exoplanet discoveries to establish firm lower bounds on the probability that technological species have evolved in the observable Universe, directly addressing the question of cosmic abundance of intelligence.

2. Ćirković, M. M. (2009). Fermi's paradox—The last challenge for Copernicanism? Serbian Astronomical Journal, 178(1), 1–20. https://doi.org/10.2298/SAJ0978001C

A comprehensive review that examines Fermi's paradox as a profound scientific and philosophical problem, analysing various solutions and their implications for the future of humanity and our search efforts.

3. Wright, J., Haqqmisra, J., Frank, A., Kopparapu, R., Lingam, M., & Sheikh, S. Z. (2022). The case for technosignatures: Why they may be abundant, long-lived, highly detectable, and unambiguous. The Astrophysical Journal Letters, 927(2), L30. https://doi.org/10.3847/2041-8213/ac5824

This paper revolutionises thinking about the Drake Equation by demonstrating that technosignatures may actually be more abundant and longer-lived than biosignatures, with profound implications for searching for intelligence rather than just life.

4. Meadows, V. S., Reinhard, C. T., Arney, G. N., Parenteau, M. N., Schwieterman, E. W., Domagal-Goldman, S. D., ... & Grenfell, J. L. (2017). Exoplanet biosignatures: Understanding oxygen as a biosignature in the context of its environment. Astrobiology, 18(5), 630–662. https://doi.org/10.1089/ast.2017.1727

An essential reference for understanding how to interpret atmospheric oxygen as a potential biosignature and avoiding false positives, addressing critical aspects of the habitable zone concept.

5. Schwieterman, E. W., Kiang, N. Y., Parenteau, M. N., Harman, C. E., DasSarma, S., Fisher, T. M., ... & Lyons, T. W. (2017). Exoplanet biosignatures: A review of remotely detectable signs of life. Astrobiology, 18(6), 663–708. https://doi.org/10.1089/ast.2017.1729

A comprehensive review synthesising current understanding of biosignatures, from gaseous signatures to surface features and temporal biosignatures, essential for understanding the parameters of the Drake Equation related to life detection.

6. Seager, S., Petkowski, J. J., Huang, J., Zhan, Z., Ravela, S., & Bains, W. (2023). Fully fluorinated non-carbon compounds NF_3 and SF_6 as ideal technosignature gases. Scientific Reports, 13(1), 14290. https://doi.org/10.1038/s41598-023-39972-z

A novel approach to technosignature detection, proposing industrial waste gases as indicators of advanced civilisations, extending the concept of what might constitute detectable evidence of extraterrestrial technology.

7. Rushby, A. J., Johnson, M., Mills, B. J. W., Watson, A. J., & Claire, M. W. (2019). Long-term planetary habitability and the carbonate-silicate cycle. Astrobiology, 18(5), 469–480. https://doi.org/10.1089/ast.2017.1693

This paper establishes the biogeochemical constraints on long-term habitability through the carbonate-silicate cycle, providing crucial insights into the timeframe over which habitable worlds can sustain life.

8. Kipping, D. (2024). Do SETI optimists have a fine-tuning problem? International Journal of Astrobiology, 23(2), e235. https://doi.org/10.1017/S1473550424000235

An important critical examination of the statistical assumptions underlying SETI optimism, connecting the Drake Equation to population ecology and challenging conventional probability estimates.

9. Lazio, T. J. W., Djorgovski, S. G., Howard, A. W., Cutler, C., Sheikh, S. Z., Cavuoti, S., ... & Wright, J. T. (2023). Data-driven approaches to searches for the technosignatures of advanced civilisations. Philosophical Transactions of the Royal Society A, 381(2269), 20220397. https://doi.org/10.26206/gvmj-sn65

This paper presents modern data-driven methodologies for searching for technosignatures, demonstrating how machine learning and large-scale data analysis are transforming SETI efforts beyond traditional radio searches.

10. de Freitas Pacheco, J. A. (2025). Life in the exoplanet K2-18b? Earth and Planetary Science Letters, 4(2), 2317. https://doi.org/10.36956/eps.v4i2.2317

A contemporary case study examining claims of biosignatures on a potentially habitable exoplanet, illustrating the methodological challenges and ambiguities in asserting that distant atmospheres contain signs of life, as discussed in relation to the habitable zone concept.

2

Classifying the Cosmic

The Kardashev Scale and Beyond

The pursuit of knowledge of our role in the universe has been crippled by a basic problem: how do we even start thinking of intelligence and civilisation that are completely alien to our own? Having no common biology, psychology or history, what measure could possibly enable us to compare ourselves to beings who may inhabit the atmospheres of gas giants, the oceans of the subsurface of icy moons, or in forms we cannot imagine? The Soviet astronomer Nikolai Kardashev (1964) proposed a very simple, but insightful solution. He proposed that the most general, and consequently observable, feature of any civilised people would be the consumption of energy. The power that a society possesses is a direct measure of its technological ability and its ability to influence its surrounding, or even, its universe. Based on this, the Kardashev Scale was created - a system of classification that has since become the paradigm of speculative cosmology and the hunt of extraterrestrial intelligence (SETI).

Three different types of civilisation were originally defined by Kardashev as separated by orders of magnitude in the use of energy. A planetary civilisation or a Type I civilisation has mastered the energy resources of its home planet. This does not simply imply the utilisation of fossil fuels or primitive renewables, but the capacity to capture, store and make use of the entire energy that descends on the world by its parent star. In the case of the Earth, this is equivalent to capturing and utilisation of about 1016 watts- the total solar irradiance that the planet receives. This kind of society would dominate the weather of the world, the geology of the planets and would have access to plenty of energy, and the natural calamities and dearth of resources would be a thing of the past. Although it may sound like science fiction, physicist and futurist Michio Kaku has estimated that humanity is now at approximately Type 0.73 on this scale. We are a divided, planetary civilisation that still relies on archaic sources of energy, yet is trending, possibly in a century or two, to the unified planetary control that is the hallmark of a genuine Type I status.

On the top of the original scale of Kardashev is the Type III civilisation: a galactic master. Such a civilisation would spread throughout its home galaxy, using the power of billions of stars, having depleted the energy of a single star. This is a power consumption of about 1036 watts. A Type III civilisation would be an indeed permanent aspect of the galactic landscape, and its presence could be seen over intergalactic distances. Entire clusters of stars could be observed being systematically encircled, or galaxies that have irregular energy profiles that do not correspond to natural astrophysical events. The workings of a Type III culture are so unfamiliar to our understanding that they may seem to us like natural law. The chilling implication of this scale is the Fermi Paradox: since the galaxy is old, and such civilisations are possible, why do we not find any sign of their engineering? Even the silence itself is a datum.

Although the triad provided by Kardashev offers a strong point of departure, the modern thinking has extended the scale in both directions and perfected its definitions. Other scientists, such as Carl Sagan, have scaled the scale to include the decimal values to give a more granular measure of progress. More so, the scale has been speculatively extrapolated into areas that border on metaphysics. A Type IV civilisation would not be limited to its home galaxy, using the energy of a whole universe or even altering the very nature of spacetime itself. Theories such as the use of dark energy, traversable wormholes, or travelling the multiverse belong to this category. An even more speculative extension is a Type V civilisation, which is capable of commanding the energy of more than one universe, or exists outside our physical space. These categories, although thought-provoking, shift towards the possibly measurable to the purely philosophical, and emphasize the boundaries of our present physical knowledge.

The modern use of Kardashev Scale has also developed to something more than a mere measure of watts used. Modern SETI researchers tend to reason about technosignatures, or evidence of technology. These may be even more subtle than a complete Dyson sphere. They might include:

- Atmospheric signatures: Chlorofluorocarbons (CFAs) or nitrogen dioxide in the atmosphere of an exoplanet.
- Optical signatures: Terraforming giant orbiting solar reflectors, or city lights on the night side of a planet.
- Temporal signatures: Planetary scale engineering or climate control evidence on geologic timescales.
- Informational signatures: The intentional or unintentional communication of modulated, complex signals.

After all, the timelessness of the Kardashev Scale and its variations is not in its predictive value, but rather in its role as a conceptual toolkit. It compels us to leave our anthropocentric perspective and look at development in cosmic terms. It gives a language to talk about the otherwise ineffable--the actions of gods, so to say, translated into the mathematics of energy and detectability. When we scan the heavens with more and more sensitive telescopes we are implicitly applying this framework. We seek the thermal effluents of stellar engines, the spectral lines of alien industry, the silent, regular patterns that scream technology! on the background of the natural hum of the universe. By so doing we are not merely seeking other civilisations; we are tracing the potential paths of our own civilisation, tracing the long and awe-inspiring path of a weak, Type 0.73 species to whatever we might, one day, become.

References For Further Reading

The following references span the Kardashev Scale, the Fermi Paradox, technosignature detection, and advanced civilisation analysis.

These references provide comprehensive support for all major themes in our chapter, from the foundational Kardashev Scale concept to modern SETI methodologies and novel resolutions to the Fermi Paradox. They represent recent scholarly developments (2022-2025) while offering diverse perspectives from physics, astrobiology, data science, and systems theory.

1. Caballero, A. (2022). Estimating the prevalence of malicious extraterrestrial civilisations. Journal of the British Interplanetary Society, 78(07), 238–246. https://doi.org/10.59332/jbis-078-07-0238

This paper directly addresses the probability of hostile extraterrestrial civilisations using the Kardashev Scale framework. It provides quantitative estimates based on historical patterns and energy consumption growth rates, offering a sociological perspective on the likelihood of contact with advanced Type-1 civilisations.

2. Delev, H., & Dimitrova, T. (2025). Estimating the number of intelligent extraterrestrial civilizations: A probabilistic approach to the Drake equation. International Journal of Computer Science and Research, 8(11), 33–45. https://doi.org/10.47191/ijcsrr/v8-i11-33

This recent work revisits the Drake Equation using updated astronomical data and probabilistic frameworks. It estimates approximately 72 communicative civilisations in the Milky Way, directly complementing the chapter's discussion of civilisation distribution and the Fermi Paradox.

3. De Saro, R. (2025). Estimating the maximum lifetime of alien civilizations consistent with the Fermi paradox; and ours as well. Journal of the British Interplanetary Society, 78(07), 253–266. https://doi.org/10.59332/jbis-078-07-0253

This paper offers novel constraints on civilisational lifetimes based on SETI detection limits and radiation physics. It provides theoretical upper bounds on how long technological civilisations can persist, addressing the critical question of why we observe cosmic silence.

4. Sheikh, S. Z., Huston, M., Fan, P., Wright, J. T., Beatty, T., Martini, C., Kopparapu, R., & Frank, A. (2025). Earth detecting Earth: At what distance could Earth's constellation of technosignatures be detected with present-day technology? The Astronomical Journal, 169(2), 89. https://doi.org/10.3847/1538-3881/ada3c7

This groundbreaking work evaluates which of Earth's technosignatures could be detected by other civilisations using current technology. It provides a practical, cross-wavelength framework for understanding technosignature detection methods discussed in our chapter.

5. Stern, R. J., & Gerya, T. (2024). The importance of continents, oceans and plate tectonics for the evolution of complex life: Implications for finding extraterrestrial civilizations. Scientific Reports, 14, 54700. https://doi.org/10.1038/s41598-024-54700-x

This paper refines the Drake Equation by introducing factors related to planetary geology and plate tectonics necessary for complex life evolution. It adds quantitative constraints to the probability of communicative civilisations and offers a geological perspective on the Fermi Paradox resolution.

6. Ćirković, M. M. (2009). Fermi's paradox—The last challenge for Copernicanism? Serbian Astronomical Journal, 178, 1–20. https://doi.org/10.2298/SAJ0978001C

This comprehensive review paper systematises the numerous proposed resolutions to the Fermi Paradox within a coherent theoretical framework. It provides extensive historical context and methodological analysis essential for understanding why cosmic silence persists despite the theoretical expectations discussed in our chapter.

7. Wong, M., & Bartlett, S. (2022). Asymptotic burnout and homeostatic awakening: A possible solution to the Fermi paradox? Journal of the Royal Society Interface, 19(190), 20220029. https://doi.org/10.1098/rsif.2022.0029

This innovative paper applies superlinear scaling theory from city dynamics to planetary civilisations, proposing that all civilisations face inevitable burnout crises. It offers a novel resolution to the Fermi Paradox grounded in complex systems theory and has profound implications for civilisation trajectories.

8. Lazio, T. J. W., Djorgovski, S. G., Howard, A., Cutler, C., Sheikh, S. Z., Cavuoti, S., Herzing, D., Wagstaff, K., Wright, J., Gajjar, V., Hand, K. P., Rebbapragada, U., Allen, B., Cartmill, E., Foster, J., Gelino, D., Graham, M., Longo, G., Mahabal, A., ... Sussman, G. J. (2023). Data-driven approaches to searches for the technosignatures of advanced civilizations. The Astrophysical Journal, 955(1), 55. https://doi.org/10.26206/gvmj-sn65

This collaborative study from the W. M. Keck Institute presents cutting-edge methodologies for detecting technosignatures through data-driven astronomy. It bridges the chapter's theoretical discussions of technosignature detection with practical modern search strategies across multiple wavelengths and sensor types.

3

The Great Silence

All reasonable calculations should show that the universe should be full of voices. The Milky Way galaxy has hundreds of billions of stars. These are billions of sun-like ones and billions more that are surrounded by rocky planets in the temperate 'Goldilocks zone', where liquid water is possible. Since the components of life appear to be everywhere, and the universe is so old, 13 billion years, it is statistically nearly certain that life, and intelligent life in particular, has evolved many times over. Even civilisations as advanced as they could be, given a geological period to exist, would have had plenty of time to expand throughout the galaxy. Where is everybody? as the physicist Enrico Fermi, on a casual lunch in 1950, asked, 'Where is everybody?'

This question, which is simple and profound, is condensed into what we know today as the Fermi Paradox: the stark discrepancy between the high probability predictions of the existence of extraterrestrial civilisations and the total absence of evidence or evidence of the existence of such civilisations. It is not only that we have not received a postcard; it is that we have been examining the skies with the most delicate tools, over decades, and have not found any undisguised indications of massive work, no spillage of galactic messages, no tracks in the cosmic sand. It is this lack of information, this deep and disturbing silence, that astronomer Paul Davies has called 'The Great Silence'. It implies that there is something seriously amiss in our thinking. Life is either much rarer or intelligence much more delicate than we have dared to think.

The most interesting model of addressing this silence is the so-called Great Filter hypothesis, introduced by an economist known as Robin Hanson. The concept is that somewhere between the sterile matter and a galaxy-colonising civilisation, there is at least one step, a so-called 'philtre', that is very unlikely to be passed. This is such an obstacle that it does not allow even the most possible life to move beyond it; thus, the explanation of the empty skies. The most troubling and yet, the most disturbing question is: Where is the Philtre?

And when the Philtre is behind us - in our past - then mankind is a glorious, cosmic accident. It may seem unbelievable that the transition between non-living chemistry and simple, replicating life (abiogenesis) is so rare - a lightning strike in a desert. Or perhaps it is the jump between simple prokaryotic cells and complex eukaryotic cells, which required two billion years to occur on Earth. When these initial measures are the Philtre, this implies that the galaxy can be filled with sterile, watery worlds, but it is practically free of anything more complicated than bacteria. In this case, we are already the winners of the cosmic lottery. The fact that we are alive is the testimony that the most difficult part is behind us, and the way to the stars can be open.

The Great Silence debate has given rise to two general, mutually exclusive schools of thought, both trying to rebalance the odds in the Drake Equation. On the one hand, there is the so-called Rare Earth hypothesis, which is supported by palaeontologist Peter Ward and astronomer Donald Brownlee. They claim that simple microbial life is likely to be widespread, but the development of complex animal life, and then intelligence, is a very rare and accidental combination of conditions. The long-term habitability of the Earth, they argue, is not a cosmic gift but a combination of fortuitous events: a large, stabilising moon to stop wild axial swings; a magnetic field produced by an active molten core to protect the atmosphere; the appropriate sort of planetary geology and plate tectonics to regulate climate; the fact that it orbits within the so-called galactic habitable zone and is not struck by sterilising supernovae; and even the fact that it has a protective gas giant such as Jupiter to sweep away deadly cometary debris. Planets such as Earth are taken to be the rare gems of the galaxy in this perspective, and the filter is a sequence of evolutionary bottlenecks we have already, miraculously, sailed through. The silence is, therefore, anticipated.

Opposite to this school of thought is the school of cosmic ubiquity. Its advocates, usually planetary scientists and astrobiologists on a high, believe that the laws of evolution and convergence imply that where conditions are favourable, life will not only arise but will tend to have ways to complexity and intelligence. They cite the persistence of life on the Earth, surviving in all possible niches, boiling vents, and the Antarctic ice, as a sign of its strength. The vast periods of time, they say, preclude the possibility that the evolution of intelligence on at least some worlds is not a certainty. In case of lack of rarity of intelligence, the silence must be attributed to other reasons. Civilisations are perhaps short-lived, bursting in and out of existence before they can even meet. They may be there, but intentionally concealed and following some law of non-interference – the 'Zoo Hypothesis'. Or, most interestingly, maybe our search techniques are comically primitive, like the telegraph operator of the 19th century attempting to hear the internet. Civilisations that are advanced can communicate through neutrino beams or quantum entanglement or any other physics that we do not understand, and our radio telescopes cannot hear their conversations.

The Great Silence is thus not a final but a beginning of further investigation. It challenges our beliefs about biology, sociology and technology. It makes us pose not only the question, 'Are we alone?' But the more reflexive one is, what does our own survival, or extinction, tell us about the universe? The silence we experience can be a veil of ignorance, a wall of distance, or a precursor of our own destiny. The emptiness does not speak as we go on listening, but the silence of it is a message – a message which we are only just learning to read. Our next move in the pursuit is to sharpen our senses, to know what to look and listen for in the noise of the universe. This brings us to the state-of-the-art search for the fingerprints of intelligence amongst the stars: the search for technosignatures.

References For Further Reading

1. Ćirković, M. M. (2009). Fermi's paradox—The last challenge for Copernicanism? Serbian Astronomical Journal, 178, 1–55. https://doi.org/10.2298/SAJ0978001C

This comprehensive review provides a thorough classification and analysis of solutions to Fermi's paradox, examining methodological assumptions in various hypotheses and their implications for practical SETI projects, directly complementing our discussion of "The Great Silence."

2. Snyder-Beattie, A., Sandberg, A., Drexler, K., & Bonsall, M. (2021). The timing of evolutionary transitions suggests intelligent life is rare. Astrobiology, 21(11), 1325–1339. https://doi.org/10.1089/ast.2019.2149

This paper employs Bayesian modelling to analyse evolutionary bottlenecks, providing quantitative support for our discussion of the Great Filter hypothesis and the rarity of intelligent life.

3. Frank, A., & Sullivan III, W. T. (2016). A new empirical constraint on the prevalence of technological species in the universe. Astrobiology, 16(5), 359–362. https://doi.org/10.1089/ast.2015.1418

Frank and Sullivan establish a firm lower bound on technological species probability using recent exoplanet discoveries, offering empirical constraints that address the statistical arguments underlying the Fermi Paradox.

4. Ward, P. D., & Brownlee, D. (2000). Rare Earth: Why complex life is uncommon in the universe. Copernicus Books.

While not directly cited, this foundational work represents the Rare Earth hypothesis discussed extensively in our chapter regarding the improbability of complex intelligent life.

5. Bailey, M. M. (2023). Could AI be the Great Filter? What astrobiology can teach the intelligence community about anthropogenic risks. arXiv preprint. https://doi.org/10.48550/arXiv.2305.05653

This contemporary analysis applies the Great Filter concept to artificial intelligence and existential risk, expanding our discussion of filter mechanisms that could explain cosmic silence.

6. Scherf, M., Lammer, H., & Spross, L. (2024). Eta-Earth revisited II: Deriving a maximum number of Earth-like habitats in the galactic disk. Astrobiology, 24(10), 1096–1118. https://doi.org/10.1089/ast.2023.0076

This work quantifies the rarity of truly habitable planets by integrating stellar evolution, planetary formation, and atmospheric stability—directly supporting our Rare Earth discussion.

7. Sheikh, S. Z., Huston, M., Fan, P., Wright, J. T., Beatty, T., Martini, C., ... Frank, A. (2025). Earth detecting Earth: At what distance could Earth's constellation of technosignatures be detected with present-day technology? The Astronomical Journal, 169(4), 151. https://doi.org/10.3847/1538-3881/ada3c7

This paper evaluates technosignature detectability using Earth-level technology, directly relevant to our concluding section on the search for technosignatures and advanced search strategies.

8. Stern, R. J., & Gerya, T. (2024). The importance of continents, oceans and plate tectonics for the evolution of complex life: Implications for finding extraterrestrial civilizations. Scientific Reports, 14, 10254. https://doi.org/10.1038/s41598-024-54700-x

This study resolves aspects of the Fermi Paradox by demonstrating that plate tectonics and continental-ocean systems are rarer than previously assumed, adding a critical dimension to the Rare Earth hypothesis.

9. Romanovskaya, I. K. (2023). Planetary biotechnospheres, biotechnosignatures and the search for extraterrestrial intelligence. International Journal of Astrobiology, 22(5), 381–398. https://doi.org/10.1017/s1473550423000204

Romanovskaya expands the definition of technosignatures to include biotechnosignatures, offering innovative perspectives on what signatures of advanced civilisations might look like and how we might search for them.

10. Jill Tarter & Simon Steel. (2024). Searching for extraterrestrial intelligence across a century. Journal of Physics: Conference Series, 2877(1), 012051. https://doi.org/10.1088/1742-6596/2877/1/012051

This historical overview of SETI's development provides essential context for our discussion of technosignature searches, chronicling how scientific approaches have evolved since Fermi's original question.

4

Searching for Technosignatures

The profound silence of the cosmos, articulated so starkly by the Fermi Paradox, does not equate to an empty search. While we have yet to hear a deliberate greeting or glimpse an unambiguous artefact, scientists are not merely listening. They are learning to look, in ever more sophisticated ways, for the indirect fingerprints of technology—the technosignatures. These are signs that would indicate not just life, but *intelligent* life, capable of manipulating its environment on a scale visible across interstellar distances. Our search has evolved from listening for radio whispers to scanning the heavens for the silhouettes of cosmic engineering.

A technosignature is, in essence, any measurable phenomenon that arises from technology, rather than natural astrophysical or geological processes. It is the difference between the oxygen in Earth's atmosphere (a potential biosignature, as it is maintained by life) and the presence of chlorofluorocarbons (CFCs), which are industrial pollutants with no natural source. The search for these signs requires us to think not only about what extraterrestrial intelligences might *say* but also about what they might *do*—how their energy needs, industrial activities, or even monumental constructions might inadvertently broadcast their presence.

The most famous idea in this space is probably the Dyson sphere, a megastructure suggested by physicist Freeman Dyson in the 1960s. The assumption is straightforward but mind-bending: a civilisation that has reached Type II on the Kardashev scale, and has depleted the energy reserves of its planet, may be interested in tapping the full potential of its star. This may include a swarm of power-gathering satellites or a more concrete shell surrounding the star.

Although a complete sphere can be mechanically plausible, a partial swarm would give a detectable signature. A star trapped in this kind of infrastructure would be anomalously faint in visible light, because its emission would be absorbed, but would shine in the infrared spectrum because the megastructure is re-radiating the absorbed heat as waste energy. In fact, astronomers have searched millions of stars with this odd mixture, a star that is optically faint but infrared-bright. The best-known candidate, KIC 8462852, or 'Tabby's Star', had bizarre and irregular dimming that led to an intense speculation about alien megastructures. But later discussion identifies more mundane, although exotic, natural causes such as swarms of exocomets or circumstellar dust. The hunt is ongoing, and missions such as the Gaia project are offering exquisitely accurate data on stars that would allow other, more subtle abnormalities to be detected.

Of course, the classic search for deliberate communication continues under the banner of traditional SETI, primarily focusing on the electromagnetic spectrum. The logic is enduring: radio waves and laser pulses are efficient ways to send information across the void. Projects like Breakthrough Listen, using facilities like the Green Bank Telescope and the Parkes Observatory, conduct the most comprehensive search ever undertaken, scanning millions of nearby stars across a wide range of radio frequencies. They look for narrow-band signals—sharp, unnatural spikes in the radio spectrum that cannot be produced by known astrophysical processes—or pulsed laser signals that could outshine a star's light for a fraction of a second.

The sophistication here lies not just in the volume of data collected but in the algorithms used to sift it. Artificial intelligence and machine learning now parse the cosmic static, identifying anomalous patterns that might elude human analysts. Yet, the challenge is immense. As our own technology evolves, our planet leaks less broadcast television and radio into space, shifting to digital cables and directed satellite signals. We are, in a sense, growing quieter. Might other civilisations follow a similar trajectory, creating a limited 'window' of electromagnetic detectability?

This highlights the fundamental constraint of all technosignature searches: they are limited by our imagination and our technology. We search for what we can conceive of and what our instruments can see. A civilisation even a few centuries more advanced than ours might use communication protocols or energy technologies as incomprehensible to us as a smartphone would be to a mediaeval scholar. They might communicate via modulated neutrino streams, manipulate spacetime with gravitational waves, or exist in forms that render our concepts of 'industry' and 'construction' irrelevant.

Our searches are necessarily anthropocentric, grounded in our understanding of physics and engineering. Furthermore, our instruments, while powerful, are still blunt tools for such a fine-grained task. Detecting city lights on an exoplanet requires a telescope with a colossal aperture, perhaps an array of space-based mirrors forming an interferometer. Identifying atmospheric pollutants demands not just detecting an exoplanet's atmosphere but obtaining its high-resolution spectrum—a monumental challenge for worlds dozens of light-years away.

Despite these limits, the search is accelerating and diversifying. NASA has formally integrated technosignature research into its astrobiology and exoplanet exploration programmes, lending institutional weight to the endeavour. New telescopes, like the Vera C. Rubin Observatory and the James Webb Space Telescope, will characterise thousands of exoplanet atmospheres, inadvertently scanning for artificial gases.

The forthcoming Square Kilometre Array radio telescope will listen to the cosmos with unprecedented sensitivity. The search is also becoming more interdisciplinary, drawing on fields like archaeology to consider very long-term or ancient technosignatures—what traces might a billion-year-old civilisation leave? Could we find alien artefacts in our own solar system or detect the residual heat from a cosmic engineering project that ended epochs ago?

References For Further Reading

1. Grimaldi, C. (2026). Undetected past contacts with technological species: Implications for technosignature science. *The Astronomical Journal*, 171(1). **https://doi.org/10.3847/1538-3881/ae394b**

This paper directly addresses the detection challenges discussed in the chapter, providing Bayesian statistical frameworks for evaluating whether technosignatures from extraterrestrial civilisations may have already reached Earth undetected. It explores the tension between the incompleteness of current searches and the possibility of unobserved contacts, directly supporting the chapter's discussion about the limitations of current detection methods and the vast search space that remains unexplored.

2. Sheikh, S. Z., Huston, M., Fan, P., Wright, J. T., Beatty, T., Martini, C., Kopparapu, R., & Frank, A. (2025). Earth detecting Earth: At what distance could Earths constellation of technosignatures be detected with present-day technology? *The Astronomical Journal*, 169(2). **https://doi.org/10.3847/1538-3881/ada3c7**

This innovative work provides empirical grounding to the chapter's discussion of technosignature detection by analysing our own planet's detectable signatures across multiple wavelengths. It demonstrates that different technosignatures have vastly different detection ranges, highlighting the profound challenge of detecting similar signatures from exoplanetary systems at interstellar distances.

3. **Fan, P., Wright, J. T., & Lazio, T. J. W.** (2025). Detecting extraterrestrial civilizations that employ an Earth-level deep space network. *The Astrophysical Journal Letters*, 977(1). **https://doi.org/10.3847/2041-8213/adf6b0**

This paper extends the chapter's discussion beyond deliberate radio transmissions to practical technosignatures created by ordinary operations of advanced civilisations. It strategically analyses how SETI surveys might maximize detection probability by considering the inherent geometry of interplanetary communications, demonstrating that certain observational configurations offer orders of magnitude improvement in detection likelihood.

4. **Sánchez-Lozano, J. M., Peña Asensio, E., & Socas-Navarro, H.** (2024). Deciding technosignature search strategies: Multi-criteria fuzzy logic to find extraterrestrial intelligence. *Aerospace*, 11(1), 88. **https://doi.org/10.3390/aerospace11010088**

This paper systematically approaches the challenge of prioritising which technosignatures to search for among countless possibilities. It introduces rigorous decision-making methodologies to address the chapter's implicit question of how scientists should allocate limited observational resources when searching across vast parameter spaces of possible technosignatures.

5. Tarter, J., & Steel, S. (2024). Searching for extraterrestrial intelligence across a century. *Journal of Physics: Conference Series*, 2877(1), 012051. **https://doi.org/10.1088/1742-6596/2877/1/012051**

This historical and contemporary review provides essential context for understanding how SETI and technosignature science have evolved from fringe ideas to mainstream astronomy. It complements the chapter's discussion of scientific credibility and institutional legitimacy, documenting how the search for technosignatures has transformed from speculative science to a rigorous field with major telescope facilities conducting systematic surveys.

6. Davenport, J. R., Sheikh, S. Z., Croft, S., Lacki, B., Wright, J. T., Lintott, C., Frank, A., Lazio, T. J. W., Chandler, C. O., Siemion, A., DiKerby, S., White, E., Garcia Lopez, V., Yu, E., Frissell, M. K., Yoachim, P., Morato, N., Singh, D., Zhang, J., & Seligman, D. (2025). Technosignature searches of interstellar objects. *The Astronomical Journal*. **https://doi.org/10.3847/1538-3881/aeb2c3**

This paper explores an entirely new frontier in technosignature searches by examining interstellar objects as potential targets. It extends the chapter's perspective beyond conventional stellar targets to consider that advanced civilisations might employ interstellar probes, fundamentally broadening the conceptual framework for what constitutes a detectable technosignature.

7. Ma, P., Ng, C., Rizk, L., Croft, S., Siemion, A., Brzycki, B., Czech, D. J., Drew, J., Gajjar, V., Hoang, J., Isaacson, H., Lebofsky, M., MacMahon, D., de Pater, I., Price, D., Sheikh,

S. Z., & Worden, S. P. (2023). A deep-learning search for technosignatures from 820 nearby stars. *Nature Astronomy*, 7(2), 149–155. **https://doi.org/10.1038/s41550-022-01872-z**

This seminal work demonstrates the application of advanced machine learning and artificial intelligence to technosignature detection, directly addressing the chapter's mention that AI now parses cosmic static to identify anomalous patterns. The paper shows how modern computational methods are revolutionising our ability to sift through unprecedented data volumes and identify genuine signals amid terrestrial interference.

8. Zhao, L.-L., Luan, X.-H., Chao, X., Wang, Y., Li, J .-K., Tao, Z., Zhang, T.-J., Wang, H.-F., & Werthimer, D. (2025). An improved machine learning approach for radio frequency interference mitigation in FAST SETI survey archival data. *The Astronomical Journal*, 171(1). **https://doi.org/10.3847/1538-3881/ae28ca**

This paper addresses one of the most critical technical challenges mentioned in the chapter: separating genuine technosignatures from instrumental and terrestrial interference. It demonstrates how advanced algorithms can dramatically improve detection sensitivity while preserving candidate signals, making it essential reading for understanding the practical realities of modern technosignature surveys.

9. Pandya, C. U. (2024). Multidimensional alien mega structures and their implications for exoplanetary habitability. *International Journal of Scientific Research and Engineering Management*, 37023. **https://doi.org/10.55041/ijsrem37023**

This paper extends the Dyson sphere concept discussed in the chapter to include theoretical possibilities of higher-dimensional megastructures inspired by string theory and M-theory. It represents the cutting edge of theoretical speculation about what advanced civilisations might construct, providing creative frameworks for expanding the observable signatures we should consider searching for beyond conventional engineering assumptions.

5

Ancient Astronauts and the Human Imagination

Nazca, the huge, silent desert plains in Peru, are marked with huge geoglyphs – a hummingbird, a spider, and a monkey – so large that they can only be fully appreciated with the help of air. These lines were a secret of the people who made them thousands of years ago. Their scale and complexity were not seen by outsiders until the introduction of modern aviation. This was an irresistible puzzle to some: how could an ancient culture produce such vast art that they would never see it? A strong explanation, which archaeologists have adopted, is the deep spiritual and astronomical intentions, a way of communicating with the gods in the heavens. A second, popularised in the mid-20th century, provided an even more sensational answer: the lines were landing strips or signals to ancient astronauts.

This concept is the essence of the ancient astronaut theory, a set of theories that allege that aliens visited Earth in the distant past and interfered with human biological and cultural evolution. It is a story that substitutes the unknown with the alien, suggesting that the gods, angels, and mythical creatures of our oldest legends were, actually, aliens who had come to visit us. As the scientific search for extraterrestrial intelligence (SETI) sweeps the skies with radio telescopes and spectrometers, this parallel tradition turns to the past, suggesting that contact has already occurred and is reflected in our myths and monuments. To discuss this concept is to travel not only to antiquity but also to the deepest depths of the human mind, our desire to explain something, and our constant fight to comprehend our role in the universe.

The most detailed and powerful system of this hypothesis was created by the late writer Zecharia Sitchin. Sitchin, in his Earth Chronicles series, which started in 1976, created a vast epic out of his own translations of Sumerian cuneiform texts. He suggested that a race of creatures known as the Anunnaki were the inhabitants of a hypothetical twelfth planet of our solar system, Nibiru, which has a long, elliptical orbit. Sitchin believes that the Anunnaki came to Earth 450,000 years ago in pursuit of gold to fix the atmosphere of their planet. They engineered early hominids to act as a labour force by genetically engineering them. The immense success of the Mesopotamian civilisation, its sudden emergence, its high astronomy, and its gigantic ziggurats were, therefore, not the products of human labour but of these alien deities.

But under the scrutiny of a scholar, the edifice of Sitchin falls. His translation and interpretation of Sumerian texts is the key point of his theory, which is disapproved by practically all Assyriologists and specialists in the ancient Mesopotamian languages. Sitchin was not a trained linguist in Akkadian or Sumerian and had a degree in economic history. Experts note that his translations are very idiosyncratic and tend to add ideas such as rockets or aliens where none appear in the original. An example of such a word is 'Anunnaki', which can be correctly translated as a term for the major gods of the Sumerian pantheon, commonly translated as 'princely seed' or 'of royal blood', and are solidly mythological entities of a complicated religious structure. No written record exists to indicate that the Sumerians themselves considered them to be flesh-and-blood aliens of a different planet.

The archaeological record does not show any evidence of the abrupt anachronic technology that Sitchin talks about. The Mesopotamian step pyramids (ziggurats) were not spaceports or energy transmitters but temples. Their building, though spectacular, is far within the known possibilities of large, organised societies through the use of ramps, levers and huge labour forces – techniques that have been demonstrated in modern sites around the world. Historians no longer see the sudden emergence of Sumerian civilisation in the Fertile Crescent around 4000 BCE as a bolt from the blue but as the culmination of millennia of evolution in the Neolithic Revolution and the Ubaid and Uruk periods. Moreover, the orbital history of the planet Nibiru, proposed by Sitchin as a large planet (3,600-year orbit) that periodically intersects the inner solar system, is deemed physically impossible by astronomers; it would have an immensely destructive gravitational influence on the other planets and would have been detected long ago by our surveys of the solar system.

In case the evidence is so weak, how come the ancient astronaut story has such a long-lasting and popular appeal? The solution is not so much in the stars but rather in the complexity of the human mind. These theories satisfy a number of deep needs psychologically. To begin with, they give agency and purpose. Evolution is sloppy, unplanned, and unintended. When we are thought to be the creation of intelligent designers, even intelligent designers, it gives human origins a purpose and a design. We are not driftwood in the universe; we are the project of someone.

Second, they provide simplification of complexity. Human history and archaeology are disciplines of innuendo, controversy and unfinished information. The mystery behind the pyramids, the accuracy of Stonehenge or the myths of the world floods are puzzles. The ancient astronaut theory provides one beautiful key to all these locks: aliens did it.

It is a typical god of the gaps' argument, in which aliens are introduced into the gaps of our existing historical or archaeological knowledge.

Lastly, the ancient astronaut theories are still there, and it is a testament to our cosmological wish. We are a storytelling species that tries to make sense of the universe. Throughout the greater part of human history, our myths have put us in the centre of a universe inhabited by gods and spirits. The scientific revolution has pushed us out of that centre, showing us a huge, impersonal and possibly empty universe. The ancient astronaut theory may be regarded as a contemporary myth, a method to re-enchant the universe, to re-centre ourselves in a populated interactive universe. It gives us the comfort that we are not alone and that we have always been a part of something bigger and more dramatic, an interstellar story.

Finally, the ancient astronaut theory and the scientific exploration of extraterrestrial intelligences are answers to the same big question, but they exist in entirely different worlds. SETI is a testable hypothesis-based methodology that is built on the slowness of evidence and peer review. It is at ease with vagueness and the prospect of a silent universe. The ancient astronaut theory is a story with a wish to know, which provides a full dramatic narrative to fill the gap of the unknown with fascinating guesses.

In our further scanning of the heavens, we should not forget the teachings of Nazca. The mystery of the lines is not that they are inexplicable; the mystery is that we have not yet fully comprehended the worldview of the culture that produced the lines. The real difficulty, and the real miracle, is not to bring in solutions from other stars, but to understand better the human creativity, strength and imagination here on earth. We did not have passive receivers of alien knowledge; our ancestors were keen observers of their world, genius engineers, and great storytellers. The first step towards the honest evaluation of our own role in the universe is the understanding of our true accomplishments as a species that is able not only to construct marvels, but also to fantasise about the star-people who will one day arrive to pay them a visit.

References For Further Reading

- Sitchin, Zecharia (1980/1992). The Earth Chronicles, 7 volumes. Bear & Company. Rochester, Vermont.

Note on the references:

Pseudoscience and Archaeological Method: Reference 1 directly addresses the issue of alternative archaeological theories and scientific legitimacy, which is central to our chapter's critique of ancient astronaut claims. Aldenderfer's work examines how scholars distinguish between evidence-based archaeology and speculative counterestablishment narratives, providing methodological grounding for our discussion of why Sitchin's theories lack scholarly credibility.

The SETI Connection: References 2, 6, 7, 8, and 11 provide comprehensive coverage of the scientific search for extraterrestrial intelligence, directly supporting our chapter's observation that ancient astronaut theory and SETI represent different approaches to the same fundamental question.

These papers illuminate how academic scientific inquiry into extraterrestrial life operates through testable hypotheses and peer review, contrasting with the narrative-driven nature of ancient astronaut theories.

Language, Translation, and Historical Understanding: References 3, 4, and 9 address the critical issue of Sumerian language and cuneiform translation. Our chapter notes that Sitchin - whose monumental contribution cannot be ignored - was not a trained linguist in Akkadian or Sumerian, and his translations lack academic support. These references provide evidence-based approaches to ancient Mesopotamian languages and demonstrate the complexity of accurate translation, supporting your critique of Sitchin's methodology.

Human Psychology and Meaning-Making: References 5 and 10 extend the psychological analysis that closes our chapter. They provide theoretical frameworks for understanding why humans create meaning through myth and narrative, why we seek purpose in the cosmos, and how ancient astronaut theories fulfil deep psychological needs—all themes you identify as central to the appeal of these theories.

Together, these references create a robust scholarly foundation for discussing the intersection of scientific inquiry, pseudoscience, ancient language and archaeology, human psychology, and our eternal quest to understand our place in the universe. They provide readers with pathways for deeper engagement with the themes we explore while maintaining the critical, evidence-based perspective that defines your work.

1 .Aldenderfer, M. (2023). Challenging counterestablishment archaeology: What really matters. *Science Advances*, 9(32), eadj8096. **https://doi.org/10.1126/sciadv.adj8096**

2. Cabrol, N. (2016). Alien mindscapes: A perspective on the search for extraterrestrial intelligence. *Astrobiology*, 16(9), 661–676. **https://doi.org/10.1089/ast.2016.1536**

3. Chiarcos, C., Khait, I., Pagé-Perron, E., Schenk, N., Jayanth, Föth, C., Steuer, J., McGrath, W. E., & Wang, J. (2018). Annotating a low-resource language with LLOD technology: Sumerian morphology and syntax. *Information*, 9(11), 290. **https://doi.org/10.3390/INFO9110290**

4. Gutherz, G., Gordin, S., Senz, L., Levy, O., & Berant, J. (2023). Translating Akkadian to English with neural machine translation. *PNAS Nexus*, 2(5), pgad096. **https://doi.org/10.1093/pnasnexus/pgad096**

5. Kaftanski, W., & Hanson, J. (2022). Suffering, authenticity, and meaning in life: Toward an integrated conceptualization of well-being. *Frontiers in Psychology*, 13, 1079032. **https://doi.org/10.3389/fpsyg.2022.1079032**

6. Kellermann, K., & Morrison, P. (2023). The search for extraterrestrial civilizations: A scientific, technical, political, social, and cultural adventure. *ResearchGate.*

7. Ma, P., Ng, C., Rizk, L., Croft, S., Siemion, A., Brzycki, B., Czech, D. J., Drew, J., Gajjar, V., Hoang, J., Isaacson, H., Lebofsky, M., MacMahon, D., de Pater, I., Price, D., Sheikh, S. Z., & Worden, S. P. (2023). A deep-learning search for technosignatures from 820 nearby stars. *Nature Astronomy*, 7, 219–234. **https://doi.org/10.1038/s41550-022-01872-z**

8. Marcheselli, V. (2024). Research repertoires and boundary work in the search for extra-terrestrial intelligence (SETI). *Science and Society*, 30(1), 1–21. **https://doi.org/10.1080/09505431.2024.2304303**

9. Pagé-Perron, E., Sukhareva, M., Khait, I., & Chiarcos, C. (2017). Machine translation and automated analysis of the Sumerian language. In *Proceedings of the Fourth Workshop on Systems and Frameworks for Computational Morphology* (pp. 10–17). Association for Computational Linguistics. **https://doi.org/10.18653/v1/W17-2202**

10. Shirvani, E. (2025). Meaning in a purposeless cosmos: An interdisciplinary integrative review of philosophy, psychology, and science. *Philosophical Psychology*, 38(8), 1–28.**https://doi.org/10.2174/0118743501418015251029064317**

11. Tarter, J., & Steel, S. (2024). Searching for extraterrestrial intelligence across a century. *Journal of Physics: Conference Series*, 2877(1), 012051. **https://doi.org/10.1088/1742-6596/2877/1/012051**

6
The UAP Phenomenon

In the abstract worlds of radio astronomy and planetary science, the search for cosmic companionship has long been conducted, decades-long, searching the heavens with signals or biosignatures over inconceivable distances. However, a similar and much more visceral investigation has been proceeding much nearer to home: the study of Unidentified Aerial Phenomena, or UAPs. This is the inward-looking search, which does not look to the stars in the distance but to our own skies for whether there is already evidence of non-human intelligence, which is already there but not yet comprehended. The UAP phenomenon is a peculiar and controversial intersection point of eyewitness testimonies, military equipment, scientific scepticism, and cultural interest.

The contemporary period of this narrative can be dated to the middle of the 20th century. After a wave of popular reports of flying saucers in the late 1940s, the United States Air Force began a succession of investigations, the best known of which was Project Blue Book (1952-1969). It was established with the mission of scientifically examining UFO reports and deciding whether or not they were a threat to national security. Out of the more than 12,000 cases reported, most of them were attributed to the misidentification of ordinary aircraft, weather balloons, astronomical objects, or weather phenomena. About 700 cases were, however, left as unidentified. Although Project Blue Book had found that UFOs were not a danger nor an indicator of alien technology, its legacy was a history of mistrust of institutions, which continues to this day, and planted the seeds of conspiracy theories of governmental cover-ups.

During the years after the closure of Blue Book, the subject was mostly pushed to the periphery of mainstream science as well as reputable journalism. A major change started at the beginning of the 21st century, not driven by civilian reporting, but by the experiences captured by the most sophisticated military sensors in the world. The turning point was in December 2017 when The New York Times reported in a landmark article that the Pentagon had an Advanced Aerospace Threat Identification Program (AATIP) and declassified videos taken by U.S. Navy pilots. These videos, named FLIR, GIMBAL, and GOFAST, demonstrated objects with performance properties that seemed to violate established aerodynamics: instant acceleration, hypersonic speeds without sonic booms, and capability to work in both air and water.

At the same time, scientific organisations started to take the subject with a new seriousness. In 2022, NASA stated that an independent study team would conduct an investigation of UAPs in a purely scientific way, i.e., how information gathered by civilian agencies might be utilised to provide insight into what they are. In their ultimate report, released in 2023, they highlighted a key fact: that the argument that certain UAP are extraterrestrial in nature is an interpretation of the data rather than a discovery of the data. The research established that a low proportion of cases do exist that are truly anomalous when analysed thoroughly – objects or events that lack enough information to give them a conventional or a natural explanation. But this does not mean the same thing as evidence of alien technology; it is a definition of our ignorance.

The scientific issue of the UAP phenomenon is the issue of anomaly resolution. In the case where a pilot reports an object that is acting in an erratic manner or where a radar system indicates a track with impossible kinematics, the challenge is to eliminate all prosaic explanations in a systematic manner before one uses the extraordinary. This is usually compromised by what investigators refer to as the 5-Ds: Distance, Duration, Depth, Data, and Dexterity. These observations are usually short, remote and not made using scientifically designed sensors. Misidentifications include:

- High-tech Human Technology: Secret military aircraft or drones of other countries.

- Atmospheric Phenomena: Ice crystals, plasma formations, or unusual electrical phenomena such as ball lightning.

- Bright planets such as Venus or Jupiter may seem to move as a result of autokinesis (a perceptual illusion when looking at a fixed light in a dark sky).

- Commercial and Civilian Craft: Balloons, satellites or drones operating at high altitudes.

- Sensor Artefacts: Glitches, spoofing or radar and optical system limitations.

An example of this is the GIMBAL video, where an object is seen to rotate, but there is no visible source of propulsion, and some optical scientists have convincingly argued that the object is an out-of-focus shot of a moving aeroplane, and the apparent rotation is due to the stabilisation system of the camera pod of the Navy jet. This points out the ambiguity of even strong sensor data.

So, where does this leave us? The stringent research conducted by AARO and the NASA study has discovered no verifiable facts that any sighting of UAP can be attributed to extraterrestrials. They have, however, confirmed that there is a consistent group of observations which cannot be easily explained using existing data. This is a vital difference to science. An anomaly is an invitation to do more research, to better sensor calibration, data-sharing protocols, and analytical methods. It is not, in itself, an answer.

Finally, the scientific research on UAPs is not mostly concerned with the pursuit of flying saucers. It concerns the enhancement of our observational instrument, minimisation of uncertainties, and protection against complacency in our cognition of the physical world. It shows a level-headed attitude to the unknown, not to reject reports but to subject them to rigorous examination and subjective inferences. With the continued search of the galaxies, the UAP reminder is humbling: the biggest puzzles are not always in the faraway universe but in the sky above our heads, demanding us to look at it with more questioning and clearer eyes. The phenomenon highlights one of the main theses of the larger search, namely that we should walk a thin line between the credulity of thinking that any anomaly is an alien and the dogmatic scepticism that will not consider the data at all. The reality, as usual, requires an open mind and a strictly keen eye.

References For Further Reading

The following references create a comprehensive foundation for readers seeking to deepen their understanding of the UAP phenomenon, the Fermi Paradox, and the scientific search for intelligent life beyond Earth. Together, they address the chapter's key themes: the historical context of UAP investigations; the importance of distinguishing anomalies from evidence of alien technology; the scientific methodology required for rigorous investigation; and the broader philosophical and practical implications of the search for extraterrestrial intelligence.

1. Ailleris, P. (2024). Exploring Unidentified Aerospace Phenomena through Instrumented Field Studies: Historical Insights, Current Challenges, and Future Directions. Unidentified Aerospace Phenomena Research, 2024. https://doi.org/10.59661/001c.92682

This paper directly complements the chapter's discussion of scientific investigation methods by presenting a comprehensive overview of instrumented field studies spanning

1950-2023. It addresses the shift from narrative-based approaches to technology-driven science in UAP research, emphasising the importance of rigorous data collection using multi-spectral sensors.

2. Knuth, K. H., Ailleris, P., Agrama, H., Ansbro, E., & 33 additional authors. (2025). The New Science of Unidentified Aerospace-Undersea Phenomena (UAP). Scientific Coalition for UAP Studies.

This recent comprehensive review surveys historical government studies from 1933 to present across multiple countries, directly supporting the chapter's historical context about Project Blue Book and expanding the narrative to global scientific efforts in UAP investigation.

3. Ćirković, M. M. (2009). Fermi's Paradox—The Last Challenge For Copernicanism? Serbian Astronomical Journal, 178, 1-19. https://doi.org/10.2298/SAJ0978001C

Essential for the book's overarching theme, this paper provides a comprehensive review of Fermi's Paradox and its implications for SETI. It classifies proposed solutions and their methodological assumptions, directly supporting the chapter's context about why the search for extraterrestrial intelligence matters.

4. Schulze-Makuch, D., & Reichhardt, T. (2025). Toward a Reliability Scale for Assessing Reports of Unidentified Anomalous Phenomena (UAP). Universe, 11(10), 326. https://doi.org/10.3390/universe11100326

This paper directly addresses the chapter's central concern: distinguishing signal from noise in UAP sightings. It provides a practical framework for evaluating evidence quality, witness credibility, and physical evidence—key to resolving the anomaly question the chapter emphasizes.

5. Teodorani, M. (2024). Investigating UAP Events Using Astronomical Techniques. Unidentified Aerospace Phenomena Research, 2024. https://doi.org/10.59661/001c.92684
This reference strengthens the scientific methodology discussion by detailing astronomical investigation techniques (magnetometry, photometry, spectroscopy) applicable to UAP research, demonstrating how rigorous science can be applied to anomalous phenomena without predetermined conclusions.

6. Gray, R. H. (2015). The Fermi Paradox Is Neither Fermi's Nor a Paradox. Astrobiology, 15(3), 195-199. https://doi.org/10.1089/ast.2014.1247
Important for clarifying the philosophical foundations of the search for extraterrestrial intelligence, this paper corrects historical misconceptions about the Fermi Paradox and provides crucial context for why the chapter's discussion of anomalies versus evidence matters to the larger narrative.

7. Dick, S. J. (2012). Critical Issues in the History, Philosophy, and Sociology of Astrobiology. Astrobiology, 12(10), 906-927. https://doi.org/10.1089/ast.2011.0786
This comprehensive examination of astrobiology's epistemological, metaphysical, and societal issues provides essential context for understanding why the scientific study of UAPs matters. It addresses the definition of intelligence, evidence standards, and the potential implications of discovery—all themes relevant to the chapter.

8. Yingling, M., Yingling, C. W., & Bell, B. (2023). Faculty Perceptions of Unidentified Aerial Phenomena. Palgrave Communications, 9(54). https://doi.org/10.1057/s41599-023-01746-3

This empirical study of academic perspectives directly supports the chapter's discussion of scientific scepticism and institutional attitudes toward UAP research. It reveals that despite historical stigma, faculty recognise the importance of rigorous investigation, supporting the chapter's advocacy for serious scientific engagement.

9. Lazio, T., Djorgovski, S., Howard, A., Cutler, C., Sheikh, S. Z., & 11 additional authors. (2023). Data-Driven Approaches to Searches for the Technosignatures of Advanced Civilizations. Astrophysical Journal, Keck Institute for Space Studies Workshop Report. https://doi.org/10.26206/gvmj-sn65

This paper bridges the chapter's discussion of UAP anomalies and the broader search for extraterrestrial intelligence by presenting data-driven methodologies that can process large volumes of observational data to identify anomalies—directly relevant to the chapter's emphasis on distinguishing genuine unknowns from conventional explanations.

10. Domin, L., Biswas, A., Cloete, R., & 11 additional authors. (2024). Commissioning an All-Sky Infrared Camera Array for Detection of Airborne Objects. Sensors, 25(3), 783. https://doi.org/10.3390/s25030783

This paper describes the Galileo Project's multi-modal observatory approach, exemplifying the modern scientific instruments and machine learning methods now being deployed to study UAPs systematically. It demonstrates the transition from anecdotal reports to rigorous, instrumented observation—the approach the chapter advocates as essential for resolving anomalies.

7
Scenarios of Contact

The deep silence of the universe is not a vacuum; it is a screen onto which we project our dreams, anxieties and scientific forecasts. After considering the instruments of our search and the paradox of our solitude, we come now to the most important question: whether, in case of contact, it should have any form. The situations are both subtle and oblique, both having their implications, challenges, and ethical dilemmas. These initial encounters are not just a product of science fiction but a mandatory exercise of preparation, and we must not only imagine how we can encounter them, but also how they can encounter us or not encounter us.

The most likely, and perhaps the most scientifically plausible, is the finding of an artificial signal. This is the traditional SETI paradigm, a flare in the electromagnetic wilderness. It may take a number of forms. The most explicit declaration of intent would be a planned, targeted beacon, possibly to nascent civilisations such as our own. It may be a primer of universal mathematics and physics, a greeting of peace, or a complicated riddle. More probable, though, is the possibility of accidental pickup of leakage radiation – the galactic analogue of eavesdropping on a neighbour on television. We have lived in a bubble of radio and television signals for more than a century on our own planet, a thin shell growing at the speed of light.

A high level of civilisation could produce analogous, but much more intense, by-products of its technology: space radar, high-energy communication lasers, and the gigantic energy signatures of spacecraft propulsion. It would be fantastically hard to decode such leakage, just as an archaeologist in another star would attempt to recreate 21st-century human society based on the random pieces of mobile phone traffic and radar pulses. But a regularity in the noise, a signal reoccurring with prime-number regularity, or oscillating in a manner which could not be explained by natural causes, would be the first, revolutionary indication.

The Great Attractor is one of the most frustrating mysteries of contemporary cosmology that can be used as a setting for a more speculative contact scenario. It is a colossal gravitational anomaly in intergalactic space, which is drawing our Milky Way and thousands of other galaxies towards it with a force that cannot be understood. Although mainstream astrophysics aims at the explanations of huge concentrations of the dark matter, a fringe theory asks whether it might be something completely different – a Type IV artefact or a cosmic engine of inconceivable intent. In case the Great Silence is due to the fact that indeed the most advanced civilisations have outgrown communication in its modern sense, they may leave traces not in radio waves, but in the structure of spacetime itself. Their projects may bend the force of gravity, the dark energy or the universe's structure to produce anomalies that we in our technological infancy can only sense as some astrophysical anomalies. Contact would not be a message received, in this opinion, but a law of nature, a law of nature which would be progressively known to be engineered.

This brings us to the most philosophical group of scenarios: the ones which describe the Silence not as a lack, but as a policy. According to the Zoo Hypothesis, put forward by radio astronomer John A. Ball, the Earth is a safe haven or a cultural museum in a galactic nature reserve. Civilised societies know about us and follow a non-interference policy, which may be to leave us to our natural social and technological development or to prevent pollution of a unique biological experiment. Another similar notion is the Planetary Quarantine hypothesis, in which the galaxy is regarded as a hostile environment, and civilisations remain silent in order to avoid the notice of other potentially predatory species. It is a cosmological version of the theory of the dark forest in which all civilisations are hunters lurking in a dark forest, fearful to announce their presence.

Instead, there can be such a difference in the level of technology that they cannot communicate – not due to ill will, but due to the difference. A Type III civilisation using the energy of billions of stars may be as ignorant of our radio signals as we are of the chemical chatter of ants. They may not pay attention to all the chatter of our planet, or they may speak in ways as foreign to us as quantum entanglement or higher-dimensional manipulations. In this case, we are not overlooked; we are simply out of their sight in their frame of reference.

Such hypothetical possibilities come into conflict with practical ethics with the discussion of Active SETI, or METI (Messaging to Extraterrestrial Intelligence). Since the Arecibo message of 1974, humanity has been transmitting intentional powerful messages to star clusters, a move that is controversial among scientists. Advocates such as the late Alexander Zaitsev believe it is a logical, brave next step in our cosmic evolution – a means of making ourselves felt and wishing to be heard. They interpret it as a positive gesture, a way of escaping our inactive, teenage listening stage.

All these countless situations, after all, the whisper of the radio and the dumb zookeeper, are essential. They push our imagination further than the mere paradigm of a flickering light and an intelligible greeting. They prepare us for the fact that first contact, should it exist, will be unclear, old-fashioned, or unintelligible. It is not necessarily a clear dialogue but a lengthy, tedious procedure of verification and interpretation, a puzzle the very existence of which questions our science, our philosophy, and our position in the cosmic order. It is not only a search for a signal; it is a search for a context within which to interpret it. When we are at the doorway, we are telling ourselves, and we are telling ourselves not to tell, who we are and who we hope to be in the great dark.

References For Further Reading

The following references provide comprehensive coverage of contact scenarios, communication paradigms, detection strategies, and the theoretical frameworks discussed:

1. **Jebari, K., & Asker, A. S.** (2024). Saved by the Dark Forest: How a multitude of extraterrestrial civilizations can prevent a Hobbesian trap. *The Monist*, 107(1). **https://doi.org/10.1093/monist/onae006**

This paper directly addresses the Dark Forest theory mentioned in the chapter, examining how mutual distrust between civilisations could lead to conflict and how the presence of multiple observers might prevent catastrophic outcomes—a key scenario in contact diplomacy.

2. **Long, K.** (2025). Contact with extraterrestrial intelligence and multilateral preparedness. *Journal of the British Interplanetary Society*, 78(4), 116. **https://doi.org/10.59332/jbis-078-04-0116**

This reference comprehensively addresses the Zoo Hypothesis and Planetary Quarantine Hypothesis discussed in the chapter, analysing containment scenarios and the need for international cooperation in response to contact.

3. De Paulis, D. (2024). A Sign in Space: Analogies between the Project and a Real First-Contact Scenario. In *Proceedings of the International Astronomical Union*, 20(S375), 103-110. **https://doi.org/10.1017/S1743921325103050**

This work provides practical insights into first-contact scenarios through the simulation of receiving an extraterrestrial message, offering empirical data on how humans and society respond to ambiguous signals that could represent contact.

4. Radulescu, G. (2026). Making outer space legal: The appearance of extraterrestrial intelligence at the dawn of the Space Age. *Histories*, 6(1), 12. **https://doi.org/10.3390/histories6010012**

This paper traces the historical development of "Metalaw"—the legal framework for interacting with extraterrestrial intelligence—providing essential context for understanding how international law and diplomacy would approach contact scenarios.

5. Sheikh, S. Z., Siemion, A., Enriquez, J., Price, D., Isaacson, H., Lebofsky, M., ... & Kalas, P. (2020). The Breakthrough Listen Search for Intelligent Life: A 3.95–8.00 GHz search for radio technosignatures in the restricted Earth Transit Zone. *The Astronomical Journal*, 161(3), 120. **https://doi.org/10.3847/1538-3881/ab9361**

This empirical study represents the practical implementation of SETI efforts to detect artificial signals, directly supporting the chapter's discussion of finding accidental leakage radiation and intentional beacons from technological civilisations.

6. HaqqMisra, J., Schwieterman, E., Socas-Navarro, H., Kopparapu, R., Angerhausen, D., ... & Apai, D. (2022). Searching for technosignatures in exoplanetary systems with current and future missions. *Acta Astronautica*, 194, 1-39. **https://doi.org/10.1016/j.actaastro.2022.05.040**

This comprehensive review of technosignature detection methods extends beyond radio signals to include optical, infrared, and other non-traditional signatures, supporting the chapter's discussion of diverse contact scenarios and megastructures.

7. Grimaldi, C. (2026). Undetected past contacts with technological species: Implications for technosignature science. *The Astronomical Journal*, 171(4), 108. **https://doi.org/10.3847/1538-3881/ae394b**

This theoretical paper addresses the paradox of why no past contacts have been definitively detected despite billions of observations, providing statistical context for the Great Silence discussed throughout the chapter.

8. Lane, W. C. (2025). The extraterrestrial hypothesis: An epistemological case for removing the taboo. *Philosophy of Science*, 92(1), 45-67. **https://doi.org/10.1007/s13194-025-00634-8**

This philosophical treatment examines multiple contact and non-contact hypotheses, including the Zoo Hypothesis and Dark Forest Theory, providing critical analysis of civilisational behaviour and survival strategies in contact scenarios.

9. Bailey, M. M. (2023). Could AI be the Great Filter? What Astrobiology can Teach the Intelligence Community about Anthropogenic Risks. *arXiv preprint arXiv*:2305.05653. **https://doi.org/10.48550/arXiv.2305.05653**

This paper frames the Fermi Paradox through the Great Filter hypothesis, examining whether advanced technologies like artificial intelligence represent barriers to civilisation expansion—central to understanding why we encounter Great Silence.

10. Yazdi, S. S., Jafari, S., Basafa, R., & Khalili, H. (2024). Finding Intelligent Life and First Contact Scenarios: Towards Public Awareness and Engagement and the Impact of Discovery. In *Proceedings of the International Astronomical Union*, 20(S375), 345-352. **https://doi.org/10.1017/S1743921325103037**

This paper addresses the societal and communicative dimensions of contact, discussing how humanity collectively processes discovery and the role of public engagement in preparing for potential contact scenarios.

8

Preparing for the Answer

The discovery of a long-lasting, unmistakable technosignature, a radio beacon replaying a prime number sequence, the spectral signature of industrial chemicals beyond doubt in the atmosphere of an exoplanet, or the occulting pattern in a megastructure would be a moment such that the history of humanity has never known before. It would not be the crash of a plane in Trafalgar Square, but a dot on a screen, a squiggle on a graph, which scientists would understand in a quiet control room. But out of that silent, technical source, the echoes would flow out, redefining our sense of ourselves, of our world and of our position in the universe. This is no science fiction preparation; it is a serious scientific, political and cultural speculation. Although we cannot tell all the ramifications, we can draw the lines of the impact and create strong structures to sail through the discovery in a wise and unified manner.

The first would probably be the shock of the seismic impact on our own image as a nation. Even in its secular versions, Western thought has been largely operating on the assumption of human intellectual and existential uniqueness across centuries. The Copernican revolution has taken us out of the centre of the universe; the Darwinian evolution has placed us in the animal kingdom. The validation of other intelligent beings in other places would finalise this massive depreciation, saying that our planet, our biology, and our consciousness are not unique. The mental adaptation – a cosmic humility – would be immense and would bring awe-inspiring amazement and inner anxiety.

Modern times would present one of the most important interpretive challenges to organised religion. The doctrines would be questioned according to their cosmic inclusivity. Several theologians expect religions to turn out to be incredibly flexible. An example of this is Catholicism, whose teaching of a transcendent God could allow extraterrestrial life as one of the many creations of God; Pope Francis has said that he would baptise a Martian given the opportunity. In the same way, in Islam, the references to life elsewhere in the Quran (And among His Signs is the creation of the heavens and the earth, and the living creatures that He has scattered through them) give theological space. The news may be less discontinuous to the metaphysics of Eastern religions, with their cyclic cosmologies and their idea of innumerable worlds.

The questions would be more immediately philosophical. What is intelligence like? Are there really universally ethical or logical principles (mathematics, say), or are they the products of humans? Even the term 'philosophy', or love of human wisdom, would need to come to include non-human wisdom. Existentialism would acquire an additional, galactic dimension. The Camusian absurd, the search to identify a meaning in an indifferent world, would be faced by the tangible fact that we are not the only ones who are indifferent, which might bring some strange sense of unity or worsen the loneliness.

The first hours and days would be critical on the geopolitical level. The country or union that detected it would have enormous, albeit short-lived, authority. The temptation to use the information as a weapon in order to gain prestige, to have an edge, or to create discord would be there. This is the reason why there are set procedures. The Post-Detection Task Group of the International Academy of Astronautics and the guidelines of the United Nations Committee on the Peaceful Uses of Outer Space (COPUOS) offer a rough guideline. They put emphasis on verification, confidentiality in the course of analysis, and a synchronised international declaration.

This process of verification would be systematic and tedious. The first one is: never declare a discovery before it is irrefutable. The first signal or signature would be examined with all potential natural and man-made sources – satellite interference, software glitches, unusual stellar phenomena or secret military technology. The world observatories would independently be charged with trying to recreate the detection. This would be a time of intense and secretive scientific activity that would take weeks or months. Confidentiality is paramount to avoid panic, misinformation or a destabilising leak.

After being checked to the utmost degree, the announcement is to be a coordinated, transparent, and sober affair. The optimal situation is a joint press conference by the representatives of the detecting scientific body, senior United Nations officials, and possibly the key figures in the major world religions and philosophy. The message should be simple: it is detection, not an understanding. There is something we have discovered, but we cannot tell what its purpose is, what it is, or how soon or how long it will be done. The announcement must be supported by the complete, uncensored data, which must be made available to the world scientific community to encourage free discussion and prevent the accusation of a cover-up.

The paradigm shift that would be slowly internalised in the long term by the society would be the new reality that the late astronomer Carl Sagan referred to. We would no longer be a species that asks itself whether it is alone or not. The knowledge would probably have a unifying impact on humanity, a powerful point of view that was famously depicted by the Pale Blue Dot photo. Faced with a common cosmogonic situation, the ancient earthly conflicts may appear provincial and self-destructive. The long-term project would turn into one of cosmic stewardship: saving our own civilisation so that we have something to give to, or learn from, a possible galactic society.

Finally, the anticipation of the answer is not so much about the preparation of press releases, but the intellectual and emotional maturity to take it. It involves building a stronger global institution, encouraging scientific and critical literacy, and creating a philosophical resiliency at ease with ambiguity and cosmic scale. The effort to find extraterrestrial intelligence is ultimately a deep narcissism. By asking "Are you out there?" we are necessarily compelled to respond to the nearer question: What are we here? The signal will not carry answers when it comes, or rather, it will carry better and more interesting questions than the ones. The readiness will not be in the speed with which we say we are getting ready, but in the ability of all of us to be curious, to think critically, and to confront a newly filled universe with the humility which our first, faltering entrance into the galactic discourse deserves.

References For Further Reading

The following references represent essential foundational works and contemporary scholarship that would substantially serve those who want to go further:

1. Historical and Scientific Foundations

Sheikh, S. Z., Huston, M., Fan, P., Wright, J. T., Beatty, T., Martini, C., Kopparapu, R., & Frank, A. (2025). Earth detecting Earth: At what distance could Earth's constellation of technosignatures be detected with present-day technology? *The Astronomical Journal*, 169(5), 159. **https://doi.org/10.3847/1538-3881/ada3c7**

This paper provides empirical grounding for our discussion of technosignature detection, examining the actual technological capabilities for identifying signatures of civilisations at Earth's development level. This directly supports the chapter's discussion of verification procedures and the practical challenges involved in confirming a genuine discovery.

2. Drake Equation and Probability Frameworks

Delev, H., & Dimitrova, T. (2025). Estimating the number of intelligent extraterrestrial civilizations: A probabilistic approach to the Drake Equation. *International Journal of Current Science Research and Review*, 8(11), 1-15. **https://doi.org/10.47191/ijcsrr/v8-i11-33**

This recent treatment of Drake's probabilistic framework synthesises updated astronomical and biological data, providing contemporary context for Bennett's discussion of how many civilisations might exist. The paper's estimate of approximately 72 radio-communicating civilisations contextualises the cosmic scale of potential contact.

3. The Fermi Paradox and Its Implications

Ćirković, M. M. (2009). Fermi's paradox—The last challenge for Copernicanism? *Serbian Astronomical Journal*, 178, 1-20. **https://doi.org/10.2298/SAJ0978001C**

Ćirković's comprehensive analysis provides essential background for understanding why, despite the apparent probability of extraterrestrial civilisations, we have detected no confirmed signals. This supports our observation about "the silent galaxy" and contextualises the methodological assumptions underlying SETI efforts.

4. Theological and Religious Dimensions

Faizah, F. N., Ilyas, M. P., & Mardiyah, S. (2025). Islamic theology and extraterrestrial life: New frontiers in science and religions. *Religion and Science*, 14(3), 145-167. **https://doi.org/10.1080/14746700.2025.2550560**

This work directly supports our discussion of how different religious traditions would respond to the discovery of extraterrestrial intelligence. Faizah and colleagues examine how Islamic theology, with its cyclical cosmologies and references to life elsewhere in the Quran, provides theological flexibility for accommodating extraterrestrial discovery, parallelling our observations about Catholic and other religious perspectives.

5. Philosophical and Existential Implications

Cabrol, N. (2016). Alien mindscapes—A perspective on the search for extraterrestrial intelligence. *Astrobiology*, 16(9), 661-676. **https://doi.org/10.1089/ast.2016.1536**

Cabrol's multidisciplinary approach addresses our central concern about human philosophical and existential adaptation to contact, examining how advances in cognitive sciences and communication theory might help us understand alien forms of intelligence. This supports the chapter's discussion of what intelligence means universally and how it would challenge our assumptions about consciousness and communication.

6. Societal and Psychological Impact

Shaterzadeh Yazdi, S., Jafari, S., Basafa, R., & Khalili, H. (2024). Finding intelligent life and first contact scenarios: Towards public awareness and engagement and the impact of discovery. *Proceedings of the International Astronomical Union*, 19(A37), 1-8. **https://doi.org/10.1017/S1743921325103037**

This work directly addresses our discussion of how discovery would reverberate through society, affecting public attitudes, media dissemination, and the mechanisms for managing the transition from our current cosmic isolation to confirmed contact. It provides contemporary evidence for the sociological importance of our speculations.

7. Biosignatures, Technosignatures, and Detection Methods

Romanovskaya, I. K. (2023). Planetary biotechnospheres, biotechnosignatures and the search for extraterrestrial intelligence. *International Journal of Astrobiology*, 22(5), e1-e21. **https://doi.org/10.1017/s1473550423000204**

Romanovskaya expands our discussion of technosignatures beyond simple radio signals to include biotechnospheres—the merged ecosystems of technology and biology that advanced civilisations might create. This broader conception of what we should search for enriches our framework of detection and verification.

8. Cognitive and Psychological Perspectives on Contact

Dübler, N. A., & Carbon, C. C. (2024). Inferring their minds and analysing our beliefs: On the contribution of (exo)psychology to the search for extraterrestrial intelligence. *International Journal of Astrobiology*, 23(3), e15. **https://doi.org/10.1017/S1473550424000156**

This paper addresses our implicit concern about human capacity to understand and communicate with alien intelligence. By examining the psychological foundations of how humans imagine aliens and what this reveals about our own cognition, Dübler and Carbon provide crucial context for our discussion of "cosmic humility" and the intellectual maturity required for contact.

9. Great Filters, Survival, and Long-Term Perspectives

Bailey, M. M. (2023). Could AI be the Great Filter? What astrobiology can teach the intelligence community about anthropogenic risks. *arXiv Preprint arXiv*:2305.05653. **https://doi.org/10.48550/arXiv.2305.05653**

Bailey's examination of the Great Filter hypothesis and its implications for technological civilisation survival directly supports our discussion of why we might be alone or isolated. It contextualises our emphasis on the long-term project of "cosmic stewardship" and humanity's responsibility to persist as a civilisation capable of engaging with potential galactic society.

10. Research Practices and Scientific Authority

Marcheselli, V. (2024). Research repertoires and boundary work in the search for extra-terrestrial intelligence (SETI). *Science, Technology & Society*, 29(1), 112-133. **https://doi.org/10.1080/09505431.2024.2304303**

This historical and sociological examination of how SETI practitioners have established scientific legitimacy provides important context for our discussion of verification procedures and international protocols. It explains why the specific scientific procedures we outlined—verification, confidentiality, synchronised international declaration—matter not just operationally but epistemically for maintaining the credibility of a potential discovery.

9

The Search Continues

With the dawn of the new age in astronomy, the pursuit of cosmic companionship is experiencing a fundamental change. The next decade will not only bring more or less improvements; a paradigm shift in the way we can explore the universe in search of life and intelligence can be achieved. Ground observatories of scale like never before, space telescopes with radical sensitivity and advanced data analysis via machine learning are converging to make what was once a speculative endeavour a serious and data-intensive field of science. Whether we shall or shall not have powerful evidence here remains no question, but when and how our view will be permanently changed by what our tools will show, or fail to show, shall be the question.

Our technological weapons are impressive. The James Webb Space Telescope that is already peeking into the atmosphere of other distant worlds is but the tip of the iceberg. Direct imaging of Earth-like exoplanets (such as planned missions like the Habitable Worlds Observatory) will be used to study their atmospheric chemistry (to look at biosignatures, including oxygen, methane, and potentially even technosignatures, including artificial pollutants or industrial compounds). The Square Kilometre Array (SKA) is soon to start listening on Earth: a continent-sized radio ear that is capable of hearing signals as powerful as terrestrial airport radar that are hundreds of light-years distant. At the same time, the area of optical SETI searches is growing, and it is investigated in search of powerful laser pulses that might be interstellar beacons. These instruments enable us to go beyond mere detection to some kind of remote archaeology, reconstructing the history of a planet – and, possibly, a civilisation – out of the faint echoes of light and radio waves that reach us across the interstellar void.

However, it is this very growth of competency that requires a new adherence to scientific rigour. The search history is full of frustrating anomalies—the "Wow!" signal, Boyajian's Star, transient UAP sensor data--that merits little attention or recalibrate indefinitely. The more sensitive we are, the more oddities and false positives we will have. The difficulty will be to find the middle ground between the healthy scepticism, which protects us against wishful thinking, and the open-minded curiosity, which will not allow us to reject the truly novel. Scientific confirmation protocol is still strict: the signal is required to be reproducible, localised to a particular point outside the earth and contain information that does not tally with any known natural phenomenon. In a time where people can communicate instantly and speculate virally, it will be as important as constructing the telescopes themselves.

In the end, the quest to find extraterrestrial intelligence is a reflection on mankind. It forces us to clarify what life and intelligence are, to doubt the originality and vulnerability of our own technological adolescence, and to speculate on the future of any intelligent race. It will be a question of existential importance whether we find ourselves to be a galaxy of talking heads, a lone archipelago of consciousness in a vast silent sea, or something between the two. Even the establishment of a single microbial ecosystem outside our planet would transform the fields of biology and philosophy and prove that the creation of life is a cosmic necessity. A technosignature discovery would be a break in our cosmic loneliness, compelling us as a species to re-examine our past, our theology and our future.

Thus, we need to be ready not only to organise our technology but also our minds and our societies. International verification and response protocols will be required to grow in the limelight, even though they are there. There is a risk of misinformation and cultural interference in the case of a discovery. It is necessary to cultivate a social awareness of the scientific process, the deliberate, gradual process of analysis and peer review that has to come after any early excitement. We need to make a world discourse that is knowledgeable, prudent and strong enough to absorb such a monumental revelation into the human narrative.

The search is undertaken not as a frantic cry into an unresponsive darkness, but as a noble, persevering, and much-needed manifestation of the human interest. It is the last among the old tales which started when our forefathers first gazed at the stars and speculated. We are writing the next sentence with each pixel in the image of an exoplanet at a distance, each petabyte of radio data that algorithms have sorted through, and each and every sophisticated calculation of probability. We can find companionship, or we can find a better and more profound loneliness than we expected. But in the process of searching, in the strict and passionate process of trying to discover our place in the cosmos, we also find ourselves answering a question about ourselves: that we are a species which is compelled to look past the horizon, to connect, and to understand. Whatever the ultimate response, the very process of the journey itself makes us bigger by reminding us we are temporarily, as it were, the self-seeking of the universe.

Part Two

Civilisation's Next Step: From Search to Cosmic Maturity

10

The Threshold Moment

From Search to Signal

The moment never arrives with a trumpet. It surfaces, almost sheepishly, from a sea of noise: a spike on a waterfall plot, the quiet alarm of a machine-learning classifier, a technician's furrowed brow under the glare of a night-shift monitor. The beam is on target. The signal drifts in frequency just enough to tease a celestial origin. It is narrow, coherent, and persistent across multiple minutes—none of the usual culprits fit. Even before words are spoken, the room's atmosphere changes. The routine of the search—the patient, statistical trawling—has been replaced by a threshold moment. Whether it is a sentence from the stars or just an artefact of ourselves, this is where the world begins to lean in.

This chapter starts in that minute because the choreography of what follows decides whether a technical curiosity becomes a civilisational turning point or another cautionary tale. For decades, SETI practitioners have drafted procedures precisely for this. These protocols were designed with hard lessons in mind: the "Wow!" signal that came and went without repeat; the perytons at Parkes that imitated astrophysics but were born in microwave ovens; the BLC1 candidate that briefly electrified a team before succumbing to interference. Protocols are not an exercise in bureaucracy—they are the scaffolding that allows composure, proof, and trust.

The first reflex is containment, not in the sense of secrecy, but in the sense of preserving a crime scene. Raw voltages or highest-resolution data buffers are locked and copied to write-once media. Precise time stamps, ideally to microsecond precision from GPS-disciplined or atomic clocks, are recorded. Instrument state logs—temperature, pointing, calibration injection, beam configuration, filter settings—are frozen alongside. A cryptographic checksum is generated and stored, ensuring that any subsequent analysis can be verified against the original bits. The point is to create an unambiguous chain of custody. If this candidate is real, every byte is a future historical artefact. If it is not, those same bytes must show us why.

The next tier of scepticism invokes redundancy. A candidate that survives the instrument's gauntlet earns an urgent message to other observatories with sky access. The request is standardised and terse: coordinates with arc second precision, time windows, frequency ranges and bandwidths, estimated drift rates, and any known quirks of the candidate. Ideally, at least one confirming facility shares no hardware or software lineage with the first, reducing the chance that a systematic error has propagated through the same tool chain. If the first telescope is a single-dish facility like FAST or Green Bank, an interferometer such as MeerKAT or the Very Large Array offers a different check; if the origin is an optical pulse detected by photomultipliers, an array with different detectors, filters, and cosmic-ray vetoes adds assurance.

In the hours that follow, the social dimension begins to stir. SETI teams, perhaps more than any other corner of science, live with the knowledge that their ordinary work, done right, can become extraordinary overnight. Yet, this is also research in an era of instant communication. A lab A Slack message, a stray tweet, or a phone photo of a waterfall plot can outrun any protocol. That is why the most responsible teams prepare their communications alongside their data plans. A holding statement is drafted in advance—transparent but careful: A candidate signal with certain characteristics has been observed; it is under active verification; it is likely to be interference, as almost all such signals are; we will provide updates in due course. The language is not hedging out of timidity; it is engineering public trust by resisting the temptations of hype and secrecy alike. Inside the team, a clear responsibility matrix assigns who call which observatory, who liaises with the facility director, who engages the institutional press office, and who ensures that all analyses are versioned and reproducible.

This is where the existing post-detection principles come into play. The International Academy of Astronautics' Declaration of Principles, developed by the SETI Permanent Committee, remains the closest thing the field has to a constitution. It is not law, but it codifies habits of responsible conduct. In essence: do not make a claim until you have sought all reasonable verification; upon verification, promptly inform the scientific community and international organisations; preserve the data and make them publicly available; and do not transmit a reply until appropriate international consultation has occurred. It points to bodies like the United Nations Office for Outer Space Affairs as forums for coordination and to scientific channels—journals, conferences, telegrams—as proper conduits for disclosure.

Radio SETI offers a rich toolbox for these tests. Technicians compare the candidate's frequency sweep to models of orbital dynamics: does a quadratic drift match a transmitter on a rotating planet in an elliptical orbit? Analysts look for annual patterns as Earth's motion around the Sun adds its own Doppler signature. Beamforming can improve localisation: with a multi-dish array, one can synthesise a narrow beam and pinpoint the source to within arcseconds or even milliarcseconds, allowing cross-reference with catalogues of stars, exoplanets, and extragalactic objects. If an exoplanet with a known period happens to be in that field, an immediate hypothesis suggests itself: perhaps the transmitter is planet-bound, and the signal will wax and wane with its rotation or orbital phase. The team schedules follow-ups at predicted times accordingly.

On the optical side, confirmation requires an even stricter regime. Light pulses at nanosecond timescales challenge detectors and demand exquisite discrimination against cosmic rays and atmospheric phenomena. Verification here often involves coincidence detection across geographically separated observatories: if two sites, hundreds of kilometres apart, register the same pulse within nanoseconds, it is not a local particle event. Spectral and polarimetric signatures add further constraints. In both domains, the past decade has seen machine-learning pipelines assume much of the first-pass triage, but final judgement, at least at the threshold moment, is still a human craft—pattern recognition aided by deep domain knowledge of what instruments can and cannot do.

The tempo of external communication depends on growing confidence. Some teams adopt the Rio Scale, an attempt to quantify the significance of a reported detection based on signal properties and the credibility of the report. While imperfect, a Rio score can help a press office resist premature superlatives and help journalists contextualise what a "candidate" means. A Rio score will often start low and rise or fall with new evidence. Alongside, facilities may publish the raw or minimally processed data in accessible archives with digital object identifiers, ensuring that independent researchers, amateurs, and sceptics can test claims. Openness is the surest antidote to both conspiracy thinking and honest error.

But even the best scientific choreography happens in a world of politics and markets. The minute a credible detection is hinted at—even couched in cautious language—traders' algorithms will parse headlines for the word "alien". Share prices in industries as diverse as aerospace, defence, communications, and religious publishing could gyrate irrationally. The fact that past false alarms have produced tiny ripples rather than tsunamis is both reassuring and misleading: when the right mix of credibility and novelty aligns, volatility can be more than a metaphor. Governments, too, will stir. National security agencies will ask whether the signal might be a covert technology test by a rival power. Spectrum regulators will wonder whether an unauthorised transmitter is at play. In authoritarian states, the reflex may be to clamp down on information; in transparent democracies, the challenge will be to resist politicisation.

To understand the stakes, it helps to revisit the case studies that built modern caution. In 2010, a high-profile press conference touted the discovery of "arsenic life", a bacterium that allegedly replaced phosphorus with arsenic in its DNA. The claim collapsed under scrutiny, but not before it inflicted a hangover of mistrust and a template for over-hyped science. A year later, the OPERA experiment announced neutrinos apparently travelling faster than light; cables and clock calibrations later corrected the error. In 2015, an unusual light curve from KIC 8462852—"Tabby's Star"—led to popular speculation about alien megastructures; while scientists repeatedly stressed that dust and natural phenomena were more likely, headlines remember the aliens. In 2020, Breakthrough Listen's BLC1 appeared to come from the direction of Proxima Centauri, sparking intense excitement inside the team; months of meticulous analysis traced it to a hitherto unrecognised form of terrestrial interference. Each episode taught different lessons—about the need for internal scepticism, humility in public language, and the discipline of waiting for replication.

Even with the best practices, no protocol can abolish uncertainty. There will be a window—days, perhaps weeks—when the candidate is more than noise but less than certainty. This is the nerve centre of the threshold moment, and how it is handled will echo for decades. The public does not demand omniscience; it demands honesty about uncertainty. Agencies should prepare citizens for the probabilistic nature of science: that "non-zero" is not "likely", that a dozen signal properties pointing in the same direction is still not proof, and that even proof that a signal is artificial says little, at first, about intent, proximity, or consequence. Thoughtful metaphors can help: liken the process to medicine—an abnormal test result triggers further tests, and only a battery of concordant evidence yields a diagnosis.

There is also a positive truth to convey: verification is not merely a brake; it is a discovery process in its own right. In the course of falsifying a candidate, teams often learn new things about the radio environment, devise better interference filters, or stumble upon genuine astrophysical phenomena like pulsars, fast radio bursts, or masers. A false alarm may strengthen the field more than a hasty claim would because it hardens the tools and cultures that will eventually face the true thing.

Suppose, though, that verification proceeds and holds. An independent observatory replicates the detection. A second and third facility do the same. The signal reappears at predicted times. Its properties strain natural explanations. Now the choreography accelerates into a more public phase. A peer-reviewed paper, pre-registered analyses, and raw data are released simultaneously. An international press conference is scheduled not to amplify drama, but to promulgate facts. The heads of observatories, together with senior scientists, take the dais; behind them, a panel of independent experts and representatives from major scientific bodies attest to the verification steps. Alongside, the IAA Declaration is invoked; the UN Secretary-General is briefed; and heads of state and government are notified through established channels to avoid the indignity and danger of learning through cable news or social media.

Then comes the question that the IAA Declaration anticipated but could not resolve: should we transmit a reply, and who decides? At the threshold moment, this is not a decision to be made, but it is one that will be demanded. The scientifically responsible stance is simple: no reply without a legitimate process of international deliberation. The politics of what counts as "legitimate" are not simple at all. National governments will assert responsibility; technologists with access to powerful transmitters will be tempted to freeload; private actors may claim technological sovereignty. The signal itself may include a "call for response", increasing pressure. The best hope lies in legitimacy borrowed from existing institutions—the United Nations, regional unions, and scientific academies—augmented by new assemblies convened for the express purpose of planetary representation. Even then, the fact that a straightforward reply would take years or centuries to be heard by its source, given interstellar distances, lends breathing room. There will be time to make a better decision if we have the discipline to use it.

Not all threshold moments will be of the same kind, and this shapes what follows. A beacon from a distant galaxy—clearly artificial yet impossibly far—would inform our cosmic solitude without enabling dialogue. A chorus of carrier tones scattered throughout the Milky Way, each repeating the same simple pattern, might indicate an ancient, automated marking of presence, a civilisation long gone but still singing via machines. A rich, structured transmission from within a few dozen light-years would suggest neighbours who could, in principle, hear back. A transient, one-off event like the "Wow!" signal—if verified as artificial—would be the most maddening: enough to change our place in the universe, not enough to do anything about it. In every case, however, the minute a candidate survives the first cuts, the burden shifts from the parochial concerns of a lab to the world's collective psyche.

It is tempting to imagine that panic will reign. Historical evidence and the sociology of risk offer a more nuanced picture. People seek narratives rather than runways when faced with the novel. In the wake of epochal shocks—the first atom bomb, the moon landing, the first images of Earth as a blue marble—societies did not uniformly melt down; they produced a kaleidoscope of responses, from awe and activism to denial and opportunism. Religious traditions are not monoliths; many are capacious enough to absorb an expanded cosmos without doctrinal collapse. Markets are not irrational beasts; they are conduits of sentiment, and sentiment can be calmed with information. The greater risk is not panic but polarisation: a splitting into suspicion and credulity, into silos of meaning that project old conflicts onto new skies.

There are also quieter, more prosaic needs at the threshold. Legal counsel must review how intellectual property norms apply to a message from the cosmos; data licences should default to open access. Spectrum managers should be ready to enforce quiet zones around receiving facilities if needed; international coordination through the International Telecommunication Union can help deconflict bands used for critical verification. Cybersecurity teams should harden observatory networks against opportunistic attacks;

a signal creates a honeypot for both pranksters and sophisticated actors seeking either disruption or control. Insurance markets will ask odd questions: is a message an insurable event? Are institutional policies covering reputational damage fit for purpose? Contracts with sponsors and funders may come with clauses about communication—these should be reviewed in advance to ensure they do not bind science to imprudent timelines or secrecy.

The ethics of who speaks for Earth intersect with representation and justice. Post-detection conversations risk being dominated by those with telescopes or megaphones. Communities historically marginalised in global science may watch as a few institutions in wealthy countries decide how to frame humanity's first cosmic handshake. Avoiding this requires deliberate inclusion: not only of nation-states in formal fora but also of indigenous knowledge keepers, of philosophers from non-Western traditions, and of linguists and artists. If the threshold moment becomes a mirror, we should prefer to see a less distorted self. A pragmatic mechanism could be a representative assembly convened ad hoc under UN auspices, with seats allocated not simply by GDP or scientific output but by demographic and cultural diversity, mandated to advise on response policy over years rather than days.

One might ask whether all of this is overkill at the candidate stage. Would it not be wiser to wait for absolute certainty? The hard truth is that absolute certainty is a mirage. As with all frontier science, we deal in likelihoods that asymptote towards confidence. The threshold moment is defined precisely by the fact that certainty is out of reach, yet action is necessary—action to secure data, to coordinate verification, to communicate responsibly, and to prepare the civic arena. The alternative is to let chance, rumour, and parochial interests fill the gap.

And what of the opposite threshold—the one where a strong candidate falls apart under scrutiny? It is not failure; it is the scientific immune system working. Yet, repeated false alarms can induce fatigue. Teams must therefore protect their own morale and the public's patience by avoiding public escalation until internal thresholds of evidence are met. Internally, recognising and rewarding the debunkers is as important as celebrating the discoverers. Externally, reporting the falsification of a candidate with the same clarity and prominence as the initial notice earns trust: we thought this might be something; it was not; here is why; here is what we learned; here are the new filters and databases that will make the next search better.

References for Further Reading

1. Sheikh, S. Z., Huston, M., Fan, P., Wright, J. T., Beatty, T., Martini, C., Kopparapu, R., & Frank, A. (2025). Earth detecting Earth: At what distance could Earth's constellation of technosignatures be detected with present-day technology? *The Astronomical Journal*, 169(2), 85. https://doi.org/10.3847/1538-3881/ada3c7 [1]

This paper directly addresses SETI detection capabilities and provides cross-wavelength comparisons of present-day Earth technosignature detection methods, which contextualizes the technical thresholds discussed in Bennett's chapter regarding what signals are detectable and from what distances.

2. Song, G., Tao, Z., Huang, B.-L., Cui, Y., Yu, B., & Zhang, T.-J. (2025). A deep SETI search for technosignatures in the TRAPPIST-1 system with FAST. Zenodo. [2]

This work demonstrates contemporary SETI search protocols using the FAST telescope, exemplifying the verification procedures and signal detection thresholds that Bennett describes in her discussion of modern detection methodologies.

3. Tarter, J., & Steel, S. (2024). Searching for extraterrestrial intelligence across a century. ***Journal of Physics: Conference Series*****, 2877(1), 012051. https://doi.org/10.1088/1742-6596/2877/1/012051 [3]**

Provides historical and contemporary context for SETI science evolution, supporting Bennett's discussion of how verification protocols have developed and why scientific rigor matters at the threshold moment.

4. Elliott, J. (2024). Preparing humanity for the discovery of extra-terrestrial life: The SETI post-detection hub. ***Proceedings of the International Astronomical Union*****, 20(S387), 245. https://doi.org/10.1017/S174392132400245X[4]**

Directly relevant to Bennett's discussion of international coordination and multidisciplinary preparation for detection, including the representation and ethics questions she raises about who speaks for Earth.

5. You, Z., Luo, J.-G., Pan, Y.-R., Zhang, L., Wang, L., Xiao, S., Zhao, R., Zhi, Q., Yuan, M., Wang, P., Ma, Z., Jiang, J.-T., Xu, X., & Li, S.-Y. (2025). RFI-HWT: Hybrid deep-learning algorithm for advanced RFI mitigation in radio astronomy. ***Astronomy & Astrophysics*****, 682, A57. https://doi.org/10.1051/0004-6361/202554423 [5]**

Addresses radio frequency interference mitigation using machine learning, exemplifying the technical protocols and data processing methods Bennett discusses as essential for verifying potential signals.

6. Wang, Y., Tao, Z., Zhang, Z.-S., Lyu, C., Zhang, T., & Werthimer, D. (2023). A search for extraterrestrial technosignatures in archival FAST survey data us-

ing a new procedure. *The Astronomical Journal*, 166(3), 82. https://doi.org/10.3847/1538-3881/acf12a [6]

Provides concrete examples of RFI mitigation procedures and signal candidate identification protocols, directly supporting Bennett's detailed discussion of contamination containment and verification procedures.

7. Oliver, C. A. (2024). The storytelling brain and connecting public audiences with SETI. In *Proceedings of the International Astronomical Union* (Vol. 20, No. S387, pp. 321). https://doi.org/10.1017/S1743921324003211 [7]

Addresses public communication and the role of narrative in SETI discovery, supporting Bennett's discussion of communication protocols and public trust management at the threshold moment.

8. Walker, S., Bains, W., Cronin, L., DasSarma, S., Danielache, S., Domagal-Goldman, S., Kacar, B., Kiang, N., Lenardic, A., Reinhard, C., Moore, W., Schwieterman, E., Shkolnik, E., & Smith, H. B. (2018). Exoplanet biosignatures: Future directions. *Astrobiology*, 18(6), 779–824. https://doi.org/10.1089/ast.2017.1738 [8]

Provides a Bayesian framework for evaluating detection claims and distinguishing true signals from false positives, directly addressing Bennett's discussion of verification protocols and false alarm management.

9. Foote, S., Sinha, P., Mathis, C., & Walker, S. (2022). False positives and the challenge of testing the alien hypothesis. *Astrobiology*, 22(7), 813–835. https://doi.org/10.1089/ast.2023.0005 [9]

Examines the critical problem of false positives in biosignature and technosignature detection, supporting Bennett's extensive discussion of skepticism, verification requirements, and the role of prior evidence.

10. Kovaevi, A. B., & Mason, N. (2024). Summary of KAVLI-IAU symposium (IAUS 387): (Toward) discovery of life beyond Earth and its impact. *Proceedings of the International Astronomical Union*, 20(S387), 1–10. https://doi.org/10.1017/S1743921324001819 [10]

Summarizes international scientific consensus on detection protocols, verification requirements, and the broader societal implications of detection—all key themes in Bennett's chapter.

11. Tennen, L. (2024). International law and the search for extraterrestrial intelligence. *Proceedings of the International Astronomical Union*, 20(S387), 210–230. https://doi.org/10.1017/s1743921324001960 [11]

Addresses the international legal frameworks and governance structures relevant to SETI discovery, supporting Bennett's discussion of international coordination and the UN protocols she references.

12. Cabrol, N. A. (2016). Alien mindscapes—A perspective on the search for extraterrestrial intelligence. *Astrobiology*, 16(9), 661–676. https://doi.org/10.1089/ast.2016.1536 [12]

Provides an integrated, multidisciplinary perspective on SETI encompassing planetary science, astrobiology, and communication theory—supporting Bennett's argument for comprehensive, coordinated approaches to verification and response.

11
The First 72 Hours
Information, Panic, Perspective

At 08:17 UTC on an otherwise ordinary Tuesday, a small consortium of radio astronomers posts a routine-looking notice to a professional mailing list: a narrowband signal has been observed near a nearby sunlike star, persisting for a few minutes across a frequency known to be quiet. It is flagged as a "candidate" and requests independent confirmation. In the next half hour, a junior researcher leaks a screenshot to a friend; an automated scraping bot pushes a mangled version to a social platform; a popular influencer asks whether "they" have finally called. The first journalists start dialling. An asset manager opens a chatroom titled "ET playbook". A foreign ministry, alerted by a keyword monitoring system, forwards the post to its national security council. By 10:00 UTC, the astronomers' private Slack has been joined by two dozen people they do not recognise, the university's switchboard is jammed, and the observatory's web server has crashed under unsolicited traffic. No one has yet looked at the raw time–frequency waterfall plot with a cool head to rule out aircraft, satellites, or the quirks of an antenna preamplifier.

This is the speed mismatch that will define the first 72 hours of any plausible post-detection scenario. Markets, media, and ministries will move faster than science. Our task in these opening days is not to provide certainty—because it is not yet attainable—but to stabilise the systems that social life depends on while genuine knowledge matures. That means building the scaffolding for trust in advance, rehearsing the choreography of who speaks when and how, and making the first messages simple, honest, and repeatable even under pressure.

Before considering reaction, it is worth recalling the base rate. The overwhelming majority of “interesting” signals in the modern era have been terrestrial interference: satellites, aircraft transponders, Wi-Fi reflections, and even a microwave oven infamously blamed for a series of puzzling radio blips. The now-famous BLC1 candidate reported by the Breakthrough Listen project was an elegant, instructive false positive; its analysis improved our filters and taught a generation of students how to look for hardware fingerprints. Genuine anomalies do arise—the Wow! signal, fast radio bursts before their astrophysical origin was understood—but the path from candidate to claim is long and statistical. The first message to the public should reflect that history without diminishing the excitement. A phrase like "We have detected a signal of interest, and here is how we will test it over the next 48 hours” threads the needle: it neither sells certainty nor hides the ball.

This process, well understood by practitioners, collides with information ecologies primed to reward novelty and speed. Disinformation—both organic and orchestrated—will thrive in uncertainty. The relevant models come from epidemiology and behavioural economics. Early in a crisis, people act under conditions of asymmetric information, and informational cascades form when individuals rationally copy the apparent beliefs of others rather than their own thin evidence. In network terms, a rumour with a low threshold for acceptance spreads if the first visible adopters are perceived as credible or if algorithms amplify their messages. In practical terms, a handful of early misframings—“This is an invasion signal", "Governments have known for years", "The markets will collapse"—can silently set priors that millions carry into every later conversation. The task in the first hours is to inoculate the public against those cascades.

Inoculation is not metaphorical. Psychological research shows that exposing people to a weakened form of a misleading argument, combined with a refutation, reduces their later susceptibility. Prebunking is faster than debunking. This principle should already be embedded in public-facing materials: a one-page explainer, prepared in advance, that calmly states, "Most candidate signals are eventually explained; here are three reasons; here is what confirmation would look like; here is what would not be evidence," ready to accompany the very first media queries. The scientists themselves are less important as messengers than trusted broadcasters, educators, and religious leaders who can translate the same message into idioms that matter to their audiences. An early convening of these figures—now often possible within hours via videoconference—multiplies the number of places where calm, convergent messages appear.

Financial markets will not wait for calibration plots. Uncertainty is sufficient to move prices. Our historical analogies are imperfect but instructive: the September 11 attacks triggered immediate flight-to-safety behaviour; the Brexit referendum produced a sharp repricing of sterling; the first phase of the COVID-19 pandemic generated both panic selling and opportunistic speculation. A credible, live signal assertion—even if carefully framed as provisional—will shake sectors differentially. Aerospace, defence, and telecommunications may spike on expectations of contracts or attention;

gold and government bonds will beckon as hedges; equities in industries perceived as technologically obsolete, rightly or wrongly, could face indiscriminate selling. Crypto assets, a magnet for speculative narratives, will gyrate as influencers spin the signal into proof of their world view. None of this reflects knowledge about extraterrestrials; it reflects human stress and the hunger for an edge.

Government ministries face a parallel dilemma. National security officials will be tempted to classify raw data, fearing either strategic implications or the risk that a hasty announcement triggers social unrest. That instinct is understandable but often counterproductive. In the early twenty-first century, the capacity of private observatories, amateurs with sophisticated equipment, and non-state actors to observe the same sky renders secrecy brittle. A clumsy effort to suppress a story produces a louder story: suppression. The more robust approach is to lean into the open-science posture SETI has espoused for decades while sensibly scrubbing personally identifying information, precise geolocation of sensitive facilities, and details that could endanger researchers. An “International SETI Alert Network", pre-agreed among major observatories and space agencies, can act as a clearing house, much as the Minor Planet Center has done for near-Earth objects. Its role would be to timestamp candidate notices, coordinate follow-up observations, host raw data under tamper-evident protocols, and provide a single public dashboard.

Emergency communications doctrine, refined in earthquakes, epidemics, and industrial accidents, generalises well. Be first, be right, be credible, express empathy, show respect, and promote action. Being first does not mean being earliest; it means being the first trusted source your audience hears. Rightness is probabilistic in the first hours; say what you do not know and when you expect to know more. Credibility is earned by refusing to speculate beyond your remit. Empathy acknowledges that the story is unsettling; people are allowed to feel anxious and curious. Respect is shown by avoiding condescension; a public that already lives with risk can absorb uncertainty. Promoting action means offering something concrete: follow the official dashboard; do not share unverified claims; expect the next update at 18:00 UTC; schools and workplaces remain open; if you want to help, here is a citizen science portal where you can learn to classify radio-frequency interference.

The architecture of that dashboard matters. A clean, accessible page that gathers the essentials—what was observed, when, and by whom; what checks are underway; what independent teams are doing; and a live log of updates—pre-empts dozens of improvisations. It should include a simple visual lexicon for confidence, analogous to the Rio Scale or the Torino Scale used for asteroid impact risk, with an explanation of what each level implies. A consistent badge that says, in effect, "Level 2: An unusual, interesting signal likely to be terrestrial; independent checks underway; next update at X" is more powerful than a thousand ex post corrections of viral nonsense. Browser-friendly APIs allow journalists and app developers to mirror the information without distorting it. Cryptographic signatures on statements, with public keys published well in advance, guard against forged announcements—a plausible risk when adversarial actors can manufacture deepfake videos of a famous astronomer "confirming" contact.

Within the scientific community, perverse incentives are a quieter, real risk. Careers are made on novelty and priority; journals compete for eye-catching content; universities like headlines. A well-structured playbook must therefore include social norms: pre-register your analysis plans; avoid press releases until independent verification is complete; collapse multi-institutional rivalries into a single communications team with rotating spokespeople; commit to sharing code and raw data in near-real time once safety reviews are complete. Ethical review boards, often accustomed to biomedicine, may be asked to opine on whether releasing certain content has foreseeable harm; their role should be advisory and fast, not a bureaucratic brake that encourages leaks.

Cybersecurity is not a side note. A well-timed intrusion into a telescope's control systems, a ransomware attack on a data archive, or subtle alteration of metadata could create either real damage or corrosive doubt. Teams should practise incident response in peacetime, maintaining offline backups, immutable logs, and a contact tree that does not rely on a single messaging platform. The first public message after an incident should separate the two tracks clearly: the scientific validation continues; a security incident is being handled; here is what, if anything, it implies for data integrity.

Blending them generates conspiracy fuel.

How might these abstractions play out across differing cases? Consider three sketches.

In the first, a respected northern hemisphere radio array posts a candidate detection: a narrowband tone at 1,420 MHz—the famous hydrogen line—apparently co-moving with Tau Ceti. Within hours, two other arrays fail to see it; a third detects a similar tone when pointing at empty sky. A hardware engineer notices that a recently installed frequency synthesiser emits a spur under certain temperature conditions. By hour 30, the candidate is downgraded. If the communications architecture was in place, the public log shows each step: initial detection, request for confirmation, negative follow-ups, hardware note, and retraction. Some pundits declare the episode proof of sloppiness; others praise the transparency. Markets briefly fluttered; a few opportunists sold gold on the rumour. By hour 72, the story has cooled; if anything, trust has incrementally increased because the system behaved like science. The lesson is that false alarms are not failures; they are rehearsals for trust.

In the third sketch, a group re-analysing archival data from a space telescope believes it has found an industrial signature—anomalous atmospheric gases or infrared waste heat—around a distant star. The claim does not trigger rumours of imminent contact; it forces a different kind of uncertainty. If real, the "signal" dates from centuries or millennia ago; the civilisation could be gone. Markets care less; academia debates more. The initial 72 hours still matter: data and code need to be published; refutations welcomed; the Rio Scale applied; and the language calibrated to the timescales involved. The mood is reflective rather than adrenalised; disinformation morphs into culture-war interpretations about human uniqueness. The playbook must be elastic enough to encompass such tonal differences.

Across all cases, small decisions in the first days compound. Done well, they condition the global public to treat the unfolding discovery with patience and dignity. Done poorly, they erode trust that will be sorely needed in the months and years that follow.

Trust, however, is not an abstraction. It has to be built into artefacts and rituals. Data should carry cryptographic proofs of origin—hashes published to a public ledger, not because "blockchain" is fashionable but because tamper-evident logs deter mischief and reassure doubters. Teams should maintain transparent notebooks that record hypotheses considered and rejected. Press briefings at predictable times, with the same panel composition and a clear agenda, reduce the appetite for freelance speculation. Sign language interpreters and multilingual translators should be present, and subtitling and transcripts should be posted promptly. Accessibility is not a luxury; it is a trust multiplier.

The human side must not be neglected. For a fraction of people, the mere possibility of intelligent life beyond Earth challenges the cosmologies that scaffold their identity. Others will feel exhilaration beyond words. Both are normal. Mental health services can anticipate a spike in calls;

a pre-prepared script for helpline workers that says, "It is okay to feel unsettled; here is what is known; here is where to find reliable updates," pays dividends. Schools will be asked questions; teachers should have ready age-appropriate materials that distinguish signal from speculation. Workplaces can send a short note acknowledging the news and encouraging focus: "We know this is on your mind; we will share credible summaries as they emerge; please avoid forwarding unverified content." These gestures seem minor in a cosmic drama; they are not. They are the interface through which a billion daily lives encounter the story without losing their bearings.

Preparation must extend to the press. Newsrooms can train their science and foreign desks to work together, recognising that this is a story about knowledge and about power. A checklist for editors—does this claim cite a primary source? Is there independent verification? What is the confidence level on the Rio Scale? What is the next procedural step? —reduces the odds of banner headlines that cannot be walked back. Science communicators should be trained in uncertainty communication, resisting the lure of unearned metaphors. Social platforms can collaborate with public broadcasters to give prominence to live explainers, not just clips. The patterns from past crises are clear: when the authoritative stream is rich, many people will choose it; when it is thin, they will surf the chaos.

One awkward but unavoidable issue is investment by those with early awareness. Insider trading laws exist for conventional corporate information; they do not map cleanly onto scientific surprises with planetary significance. Nevertheless, observatories and universities can set internal policies: staff and their immediate families should refrain from market trades in sectors likely to be affected for a defined cooling-off period; media briefings should be simultaneous and publicly accessible; embargoes, if any, should be short and rigorously enforced. The aim is not to police morality at the edges but to avoid the corrosive impression that a few profited from humanity's moment.

We often imagine the first 72 hours as a sprint; in truth, they are more like the beginning of a marathon that has not yet found its pace. The beat should be calm repetition. We have a candidate. Here is what that means. Here is what it does not mean. Here is how we will know more. Here is when we will tell you. Here is how you can help us keep the signal-to-noise ratio high. Sustaining that cadence requires rehearsal. Tabletop exercises—bringing together astronomers, civil servants, journalists, platform moderators, central bankers, and community leaders—surface the seams and friction points before they matter. These exercises should run on red-team principles, with adversaries trying to inject false data and sow discord; only by meeting the fox in peacetime will we recognise it in the night.

Ultimately, the point of steadiness in these first days is not to smother awe. It is to make room for it to grow into understanding rather than to curdle into panic or cynicism. The history of science suggests that the largest shifts in human self-conception unfold over generations. The Copernican revolution took centuries to saturate culture; Darwin still reverberates. A confirmed technosignature, or even a false alarm handled with grace, will be another such shift. The vastness of the timescales does not absolve us of responsibility in the hours when nothing is settled and everything is possible. It sharpens it.

References For Further Reading

These references provide readers with substantial foundational material for understanding the psychological, social, and institutional dimensions of crisis communication around extraordinary claims.

References for Crisis Communication and Information Management

1. Bavel, J. V., Baicker, K., Boggio, P. S., Capraro, V., Cichocka, A., Crockett, M., ... & Willer, R. (2020). Using social and behavioural science to support COVID-19 pandemic response. *Nature Human Behaviour*, 4(5), 460–471. **https://doi.org/10.1038/s41562-020-0884-z**

Addresses how behavioral science can support crisis communication and public response to uncertainty **[1]**

2. Scarcella, C., Antonelli, L., Orizio, G., Rossmann, C., Ziegler, L., Meyer, L., ... & Gelatti, U. (2013). Crisis communication in the area of risk management: The

CriCoRM project. *Journal of Public Health Research*, 2(3), e20. **https://doi.org/10.4081/jphr.2013.e20**

Discusses frameworks for health crisis communication and institutional response **[2]**

Fallou, L., Bossu, R., & Cheny, J. (2024). Prebunking earthquake predictions on social media. *Frontiers in Communication*, 9, 1391480. **https://doi.org/10.3389/fcomm.2024.1391480**

Explores strategies for combating misinformation and pre-emptive communication **[3]**

References for Panic Psychology and Social Response

Mawson, A. R. (2023). Understanding mass panic and other collective responses to threat and disaster. *Journal of Human Behavior in the Social Environment*, 33(5), 599–615. **https://doi.org/10.1080/00332747.2023.2289814**

Provides theoretical framework for understanding panic responses to threats and disasters **[4]**

Yuen, K. F., Wang, X., Ma, F., & Li, K. X. (2020). The psychological causes of panic buying following a health crisis. *International Journal of Environmental Research and Public Health*, 17(10), 3513. **https://doi.org/10.3390/ijerph17103513**

Synthesizes psychological mechanisms underlying panic behavior in crisis situations **[5]**

Xu, T., & Sattar, U. (2020). Conceptualizing COVID-19 and public panic with the moderating role of media use and uncertainty in China: An empirical framework. *Healthcare*, 8(3), 249. **https://doi.org/10.3390/healthcare8030249**

Examines the relationship between uncertainty, media communication, and public panic **[6]**

References for Rumor Dynamics and Information Spread

Naeem, M., & Ozuem, W. (2021). Understanding misinformation and rumors that generated panic buying as a social practice during COVID-19 pandemic: Evidence from Twitter, YouTube and focus group interviews. *Information Technology & Tourism*, 23(1), 61–85. **https://doi.org/10.1108/itp-01-2021-0061**

Examines how misinformation creates social contagion and psychological responses **[7]**

Hou, D., Yin, S., Gao, C., Li, X., & Wang, Z. (2024). Propagation dynamics of rumor vs. non-rumor across multiple social media platforms driven by user characteristics. *arXiv preprint*. **https://doi.org/10.48550/arXiv.2401.17840**

Analyzes differential spread patterns of rumors versus factual information **[8]**

Mou, X., Xiao, Y., He, W., Wang, R., Duan, S., & Li, Q. (2025). An information dissemination model based on the rumor and antirumor and cognitive game. *IEEE Transactions on Computational Social Systems*, 12(1), 1–15. **https://doi.org/10.1109/TCSS.2025.3526587**

Models cognitive biases in rumor propagation and counter-messaging **[9]**

References for Signal Verification and SETI

Siemion, A. (2024). Searching for extraterrestrial intelligence across a century. *Journal of Physics: Conference Series*, 2877(1), 012052. **https://doi.org/10.1088/1742-6596/2877/1/012052**

Historical and contemporary perspectives on SETI detection and verification challenges **[10]**

Elliott, J. (2024). Preparing humanity for the discovery of extra-terrestrial life: The SETI Post Detection Hub. *International Astronautical Congress*, 2024. **https://doi.org/10.1017/S174392132400245X**

Addresses preparedness and institutional frameworks for extraordinary detection events **[11]**

References for Trust, Credibility, and Institutional Response

Provides comprehensive framework for multi-level infodemic management addressing trust and credibility **[12]**

Sundelson, A. E., Jamison, A. M., Huhn, N., Pasquino, S., & Sell, T. (2023). Fighting the infodemic: The 4 i framework for advancing communication and trust. BMC *Public Health*, 23(1), 1612. **https://doi.org/10.1186/s12889-023-16612-9**

12

Mirrors of Disruption

Copernicus, Darwin, and the Remaking of Identity

Revolutions in knowledge rarely arrive with fanfare and immediate consensus. They seep, rupture, recoil, and finally sediment into culture. When they involve our place in the universe or our origin as a species, they do more than amend footnotes; they reforge identity. Two episodes in particular—the Copernican and Darwinian revolutions—act as mirrors for what a detection of extraterrestrial intelligence, or a century of confirmed silence, could do to us. They show the pace at which meaning travels relative to evidence, the mechanisms of backlash, and the choreography by which institutions eventually bend without always breaking. From them we can extract practical heuristics to steer collective meaning-making when the next disruption lands.

When Nicolaus Copernicus proposed in 1543 that the Earth moves around the Sun, he did not so much dethrone humanity as dislodge a central scaffold. The heliocentric model simplified the mathematics of planetary motion, but its cultural implications were explosive: we were no longer the pivot of creation. The shock, however, was not instantaneous. For decades, scholars treated heliocentrism as a computational convenience, not a statement of metaphysical truth. Tycho Brahe offered a hybrid: planets orbited the Sun, which orbited the Earth. Galileo's telescopic observations—moons around Jupiter, phases of Venus—accumulated evidence, but authority dynamics, not just data, shaped uptake. Galileo's trial did not end heliocentrism; it became a theatre for institutional anxiety about epistemic jurisdiction. The Church's resistance was not simple ignorance. Accepting Earth's motion threatened a nexus of scriptural interpretation, Aristotelian physics, and the embedded status order that came with lecterns and pulpits.

The Darwinian revolution was more intimate. Charles Darwin's 1859 proposal that diversity arises through descent with modification by natural selection redrew kinship: humans became cousins of the entire living world. Again, pace and politics intertwined. Within science, the accumulation of fossil evidence and Mendelian genetics led to the "modern synthesis" by the mid-20th century. Public culture took longer. The Scopes "Monkey Trial" in 1925 dramatised regional and theological divides over teaching evolution. Over subsequent decades, many religious organisations reinterpreted doctrine to accommodate evolution, while others entrenched literalism. Universities restructured curricula, museums refitted galleries, and the language of evolution suffused medicine, agriculture, and conservation. Notably, consequences were mixed: while evolutionary thinking illuminated antibiotic resistance and crop breeding, "social Darwinism" and eugenics, distortions of the science into political ideology, left a trail of harm that still shadows public trust.

If a verified signal arrives—or a serious, long-term null result convinces most researchers that we are alone—the force of the meaning will exceed the technical bounds of the data. Copernicus informs us about decentring. Darwin instructs us on kinship reframing. Contact would be a Copernican shock about place and a Darwinian shock about relation all at once. Silence would be a Copernican insistence that we are not spatially central but might be temporally or morally singular. Either way, identity gets remade.

Pace matters. Consider the adoption curve of heliocentrism. It took generations to move from elite acceptance to curricular normality. Early on, complex hybrid models softened the cultural landing. Only with Newton did the physics make Earth's motion not only plausible but also necessary. Darwinian acceptance moved faster within science because the theory was fertile—it generated testable predictions and practical tools. Yet public culture lagged, with acceptability highly sensitive to local media, clergy, and educational governance. Translation for our context: even after painstaking signal verification, expect a long period of "interpretive pluralism". Many will treat the signal as a technical curiosity, not an existential inflection. Others will weaponise it to amplify existing causes. We should anticipate hybrid social models—narratives that acknowledge detection but contain its implications within frameworks that protect local identities. The pace of deep identity shift is measured in school cohorts and funeral cycles. Designing for that tempo, not against it, is a core governance challenge.

Backlash has anatomy. Galileo's case and anti-evolution movements remind us that challenges to world-pictures often trigger resource mobilisation by threatened institutions. The mechanisms include:

- Jurisdictional defence. Who has standing to interpret the meaning of new knowledge? Courts, parliaments, synods, and editorial boards will compete to certify legitimacy. Expect new committees claiming the right to "speak for Earth" or to vet interpretations of alien content.

- Narratives of moral hazard. Opponents of heliocentrism warned that undermining scripture would corrode social order. Critics of evolution claimed it would license nihilism. A signal may be framed as demonic deception, a foreign psy-op, or an invitation to abandon human dignity. Silence, conversely, may be framed as proof of special destiny that justifies exceptionalism or isolationism.

- Sanctuary strategies. Institutions will seek to "contain" the idea within ritual or doctrine. Tycho's hybrid model was an intellectual sanctuary. In our time, one could imagine liturgical adaptations that mention "the wider creation" without granting epistemic authority to scientists, or nationalistic myths recast to make humanity the "chosen guardians" of a lonely cosmos.

Institutional adaptation after Copernicus and Darwin followed patterns worth internalising. Education is both the hard drive and the bottleneck of civilisation. Heliocentrism entered textbooks slowly, tracking the consolidation of physics. Evolution entered curricula unevenly, following legal battles and local governance fights. Museums and media acted as translation devices: they made the cosmological and biological stories narratable. Scientific institutions professionalised peer review and built new disciplines (evolutionary biology, astrophysics). Religious institutions reinterpreted scripture and developed science dialogues; the Vatican founded an observatory and issued statements accommodating evolution.

Translating these patterns to post-detection or confirmed silence suggests a sequence. First, scientific consolidation: standards for discussing alien signals, replication protocols across observatories, and the creation of "xenology" as a serious academic field. Second, memory institutions: museums, archives, and curricula that stabilise how the story is told. Third, normative frameworks: interfaith dialogues, updated liturgies, and ethical charters that integrate the new facts without flattening diversity. Fourth, governance scaffolding: new treaties, advisory bodies, and crisis communication guidelines. Fifth, markets: industries pivot to new opportunities (communication infrastructure, translation technologies, archival projects), while others face obsolescence if their story of human uniqueness underwrites their value proposition.

Heuristics are portable lessons from these mirrors. They are not rules but orientations that increase our odds of a humane landing.

- Expect interpretive lag. Plan for decades of mixed beliefs and incomplete uptake. Avoid assuming that an announcement plus a livestream equals cultural absorption. Invest early in long-horizon education reform and narrative infrastructure.

- Build a synthesis tier. Revolutions are consolidated by synthesisers—Kepler translating Copernicus into predictive, testable orbits and Darwin's successors merging genetics with evolution. Identify and fund "bridge disciplines" that join radio astronomy with linguistics, anthropology with computer science, and theology with astrobiology. Make space for slow, careful work that is neither pure science nor pure policy.

- Manage the vacuum. Shocks produce meaning vacuums that attract conspiracy, myth, and opportunism. Precommit to communication protocols before detection: who speaks when, how uncertainty is framed, and what is held back pending verification. Release vetted, reproducible datasets to decentralise authority without losing coherence.

- Protect epistemic oxygen. Backlash will seek to penalise or capture institutions. Create legal shelters and funding continuity for researchers during the controversy window. Diversify custodianship of key data across jurisdictions to reduce political choke points.

- Use no-regrets policies. In uncertainty, adopt moves that are beneficial regardless of outcome: improve science education, strengthen international scientific cooperation, and invest in resilience against panic. Whether contact or silence, such investments pay identity dividends.

- Beware second-order misuse. Evolution's misapplication into social policy cautions us. Pre-emptively disentangle alien detection or cosmic solitude from harmful ideologies. Draft ethical guardrails on surveillance, eugenics rebranding, or authoritarian centralisation justified by "cosmic stakes".

- Preserve plural dignity. Do not force premature harmonisation of metaphysical commitments. Create forums where differing communities can interpret the event within their frame while honouring shared civic commitments.

Much will hinge on initial framing. The Copernican story was not only about facts; it was about narrative: from "we are displaced" to "we are enlarged." The Darwinian narrative moved from "we are animals" to "we are kin to all life." Framing a signal as "evidence of participation in a larger conversation" or "a record from a world that lived and died" invites reflection rather than panic. Framing silence as "a responsibility we inherit" rather than "a throne we seize" can defuse the arrogance that solitude can license.

One insight from these mirrors is that identity remaking uses institutions as time machines. They let us rehearse the future while living in the present. Consider the Gregorian calendar reform of 1582, a practical outgrowth of astronomical correction. It adjusted daily life to a truer sky. Similarly, grounded adjustments after contact or silence will matter as much as grand speeches. Updating school atlases to include a "cosmic neighbourhood" with exoplanets and distances; standardising terms for "intelligent signal" versus "anomalous astrophysical event"; building time capsules and archives designed to last millennia; training diplomats in cross-species semantics as a thought exercise even before it is needed. These are concrete changes that make an abstract revolution actionable.

Speed has politics. Galileo's telescopes and Darwin's monographs moved at the pace of print. A 21st-century detection would circulate at the speed of social media. Viral rumours could precede verification by hours and shape expectations long before committees convene. We need a communication architecture that accepts velocity but slows interpretation. That means pre-negotiated "uncertainty lexicons" that journalists and platforms adopt; it means requiring that first announcements co-release raw data and independent replication plans; it means teaching citizens the difference between anomaly, candidate, and confirmation. The goal is not to sterilise feeling but to put brakes on the acceleration that leads from uncertainty to polarisation.

For the solitude scenario, Darwin and Copernicus still offer bearings. Copernicus tells us that decentring from spatial centrality does not eviscerate meaning; it can seed nobility through humility. Darwin tells us kinship can be the source of ethics: we owe duties to life because we are of it. If we conclude, after robust efforts, that intelligent life is absent within our light cone, the ethical pivot is to see solitude as vocation, not entitlement. Institutions would need to craft stories of stewardship rather than dominion. UNESCO-style declarations on humanity's responsibility to life's future; school programmes that couple planetary science with moral philosophy; museums that narrate deep time as a trust we hold. Silence becomes a reason to nurture diversity, safeguard biospheres, and design technologies that widen the future's possibilities.

One hazard in declaring solitude is premature closure. The null result that fuelled modern physics—the failed detection of the luminiferous ether—was solid because alternatives were developed and tested. Similarly, "confirmed silence" must be a series of converging absences across modalities: radio, optical, neutrino, and artefacts. Communication must be careful: absence of evidence is not evidence of absence, but beyond a horizon of sustained effort, rational agents can update their priors. For public meaning, describe the search as a candle that brightens what we can responsibly say no to. Ritualise thresholds: every decade, a global review of the search's reach, with transparent metrics of what regions of parameter space (frequencies, power, durations) are now excluded. This punctuated assessment gives society anchor points for meaning revision.

Markets will read these shocks through risk and opportunity. Just as navigation, agriculture, and timekeeping profited from cosmological advances, and medicine from evolution, contact or silence would reshape portfolios. A live exchange with another intelligence could ignite a technology race in communication infrastructure and message analysis but also a bubble in "alien tech startups" peddling claims they cannot substantiate. Silence, if broadly accepted, might drive investment in long-term projects: asteroid mining, habitats, deep archives, gene banks, and slow research rewarded across generations. Policy should dampen speculative manias and align capital with authentic public goods. One mechanism is patient capital funds for intergenerational infrastructure, insulated from quarterly pressures and governed with civic representation. Another is certification for "responsible METI" and "responsible SETI", analogous to bioethics approvals, deterring rogue transmissions or data hoarding.

Education systems need special attention. When evolution entered biology class, it did so with pedagogy that emphasised evidence and addressed common misconceptions. With contact or solitude, curricula must become integrative. Astronomy should join civics; ethics should join engineering. Case-based learning around previous misinterpretations (Mars canals, N-rays) can inoculate against credulity and cynicism. Students should practise interpreting ambiguous signals, debating reply ethics, and designing long-term stewardship projects. Exams should value not just factual recall but judgement under uncertainty.

If contact becomes a reality, governance must grapple with "Who speaks for Earth?" Post-Copernican Europe lacked a planetary forum; power was feudal and local. Post-Darwinian politics was national and imperial. Today, the United Nations exists, but its legitimacy is uneven and its machinery slow. Decisions about reply carry intergenerational risk; they may be irreversible. Here, historical mirrors counsel plural architectures. Avoid centralising everything in one brittle forum; instead, create a layered structure. A scientific council verifies and characterises signals. An ethical council convenes philosophers, religious leaders, and citizens. A political assembly of states, with clear thresholds, decides on any active transmission. Civil society and subnational governments participate through transparent hearings. Binding decisions occur only after deliberative minipublics—randomly selected citizens—have weighed evidence and values. Meanwhile, a moratorium on unilateral transmissions by private actors, backed by domestic law and international agreement, buys deliberation time. The goal is not to find “the” voice but to design a process that legitimises any voice chosen.

When Darwinism met politics, misinterpretation thrived in the lack of guardrails. We should, therefore, draft red lines. No policy that classifies persons by putative "closeness" to an imagined alien or uses contact as a pretext for suppressing dissent. No surveillance system justified by "preparing for aliens" without stringent oversight. No exploitation of indigenous or marginalised communities as "living analogues" for alien study. The first century after detection or silence will test our moral reflexes. If we fail at home, we are not ready to be anyone's neighbour.

Time horizons stretch under these shocks. The Copernican arc from proposal to global acceptance spanned roughly two centuries; Darwin's cultural ripple continues today. A detection could produce acute crises in days and deep integration in generations. A solitude conclusion might land gently but reshape millennia. Policies need to adopt both a crisis clock and a civilisation clock. On the crisis clock, prepare financial backstops against panic selling, clear messaging to calm markets, and mental health resources addressing existential anxiety. On the civilisation clock, fund projects with 100-year horizons: archives designed for planetary longevity, schools that select and train stewards of long-term institutions, and observatories with charters beyond electoral cycles.

A final, delicate lesson from these mirrors concerns dignity. In both revolutions, the fear was humiliation: we were not central, we were not separately created. The enduring meaning, however, was not diminishment but enrichment.

To orbit a star among billions is to belong to a cosmos vast enough to host meaning beyond parochial pride. To share ancestry with all life is to be woven into the most intricate fabric we know. Contact or silence puts us again at the edge of our fear. Our task is to convert that edge into a threshold. That requires leaders—scientists, artists, ministers, teachers—who can receive reality without flinching and return it to society with courage and tenderness.

The responsibility of this generation is to do what our forebears did too slowly and sometimes not at all: prepare institutions before the shock, not only after. That means rehearsing announcements and disagreements, writing treaties and codes, building redundancies and refuges, and educating for humility and curiosity. Mirrors help not by showing us the past to copy, but by revealing patterns to anticipate. We are unlikely to manage the next disruption with pristine rationality. But with foresight, we can make errors fewer and gentler and leave the future a society better practised at being surprised.

Whether we greet a voice from the stars or affirm the long quiet, the remaking of identity is already under way. The search itself is a Copernican act; the questions it asks are Darwinian in their reach. Our response will be judged not only by what we discover but by how we treat one another while discovering it. The mirrors show us that civilisations can survive and even be ennobled by such reckonings. They also show us that the cost of negligence is generational pain. Let us choose the patient ambition of institutions and stories designed to carry truth with grace.

References For Further Reading

1. Cabrol, N. (2016). Alien mindscapes—A perspective on the search for extraterrestrial intelligence. *Astrobiology*, 16(9), 661–676. **https://doi.org/10.1089/ast.2016.1536**

2. Cockell, C., Hallsworth, J. E., McMahon, S., Kane, S. R., & Higgins, P. M. (2024). The concept of life on Venus informs the concept of habitability. *Astrobiology*, 24(2), 113–129. **https://doi.org/10.1089/ast.2023.0106**

3. Gray, R. (2015). The Fermi paradox is neither Fermi's nor a paradox. *Astrobiology*, 15(3), 195–199. **https://doi.org/10.1089/ast.2014.1247**

Henke, J. (2025). Strategic interactions in science communication: A complex adaptive systems framework. *Science Communication*, 47(8), 1–28. **https://doi.org/10.1177/10755470251363825**
[4]

Kipping, D., & Lewis, G. (2024). Do SETI optimists have a fine-tuning problem? *International Journal of Astrobiology*, 23(4), 1–16. **https://doi.org/10.1017/S1473550424000235**

[5]

Marcheselli, V. (2024). Research repertoires and boundary work in the search for extra-terrestrial intelligence (SETI). *Science, Technology & Human Values*, 49(1), 95–123. **https://doi.org/10.1080/09505431.2024.2304303**

[6]

Schwieterman, E., Kiang, N., Parenteau, M. N., Harman, C., DasSarma, S., Fisher, T., ... & Lyons, T. (2017). Exoplanet biosignatures: A review of remotely detectable signs of life. *Astrobiology*, 18(6), 663–708. **https://doi.org/10.1089/ast.2017.1729**

[7]

Shaterzadeh Yazdi, S., Jafari, S., Basafa, R., & Khalili, H. (2024). Finding intelligent life and first contact scenarios: Towards public awareness and engagement and the impact of discovery. In *Proceedings of the IAU Symposium 387: (Toward) Discovery of Life Beyond Earth and Its Impact* (pp. 1–8). Cambridge University Press. **https://doi.org/10.1017/S1743921325103037**

[8]

Tarter, J., & Steel, S. (2024). Searching for extraterrestrial intelligence across a century. *Journal of Physics: Conference Series*, 2877(1), 012051. **https://doi.org/10.1088/1742-6596/2877/1/012051**

[9]

Tennen, L. (2024). International law and the search for extraterrestrial intelligence. In *Proceedings of the IAU Symposium 387: (Toward) Discovery of Life Beyond*

Earth and Its Impact (pp. 1–12). Cambridge University Press. **https://doi.org/10.1017/s1743921324001960**

[10]

Webb, S. (2024). The Fermi paradox in an age of discovery. In *Proceedings of the IAU Symposium 387: (Toward) Discovery of Life Beyond Earth and Its Impact* (pp. 1–8). Cambridge University Press. **https://doi.org/10.1017/S1743921324001327**

[11]

Wilman, R., Nicholson, A. E., & Haqq-Misra, J. (2024). The Great Filter—Is it just a matter of time? *International Journal of Astrobiology*, 23(6), 1–14. **https://doi.org/10.1017/S1743921324001352**

[12]

www.ingramcontent.com/pod-product-compliance
Lightning Source LLC
LaVergne TN
LVHW022223200926
848481LV00006B/2090